For Leslie.

H_2O: The Guide to Quality Bottled Water

Arthur von Wiesenberger

Woodbridge Press
Santa Barbara

This book has evolved from ancient and modern writings, scientific findings, newspaper reports, lectures, personal interviews and conventions. I make no claims that weight reduction, improvement of appearance, or cures for any physical or mental ailments will result from drinking water.

—ARTHUR VON WIESENBERGER

Published and distributed by

Woodbridge Press Publishing Company
Post Office Box 6189
Santa Barbara, CA 93160
815 De La Vina Street
Santa Barbara, CA 93101

Printed in the United States of America.

Distributed simultaneously in Canada.

Library of Congress Cataloging-in-Publication Data

von Wiesenberger, Arthur.
H_2O: the guide to quality bottled water.

Bibliography: p.
Includes index.
1. Drinking water, Bottled. I. Title.
TP620.V67 1988 663'.61 88-5699
ISBN 0-88007-166-4

Acknowledgements

Bottled water is an industry which brings together people who, as a group, seem to have a higher goal. Although there is great financial reward in water, there is also a common current toward health, purity and quality. Water, by its unique mystical and essential role in our lives, touches those who study it for science, profit or pure fascination. I have been fortunate to meet many of those individuals along the path of writing H_2O. From spring owners to philosophers, from scientists to entrepreneurs, each has added greatly to this book.

Leading my acknowledgements is one to Anne Lowenkopf whose unending enthusiasm and personal water knowledge have helped to make H_2O a reality. John Scott of Mountain Valley Water has given much guidance over the years, and William Deal of the International Bottled Water Association has been very helpful. I also want to thank Ron Davis and Jane Lazgin of The Perrier Group; Jack Maguire, Rene Antoine, Vincent Ducasse and Thirry Le Duc of Evian; Guy de la Motte-Bouloumie and Philippe Lecas of Vittel; Dr. Marino Viola at San Pellegrino; August Busch III, Thomas Aldrich and Mike Richter at Anheuser-Busch Companies, Inc.; George J. Schmitt of Hinckley & Schmitt, Inc.; Marcus Bender of Hawaiian Water Partners; Rick Scoville of Artesia; Austen and Richard Hess of Ephrata Diamond Spring; Fred Sipper of M&M Distributing; Rick Sajbel at Manitou Corporation; Dr. Jack M. Sheneman at the California Department of Health Services; Thomas C. Bell at *U.S. Water News;* Julia Child; Robert Mondavi; Wolfgang Puck; Claudia Stone; William Tomicki; Pamela Lechtman; Carol Buteau; Howard Weeks; and Noel Young.

The men and women that work with the bottled water companies listed in Part Two have all added to this book and space does not allow acknowledging them in detail. However, to those who have given me suggestions and inspiration, from home delivery man to corporate president, thank you. For any excellence in this volume, these men and women share the credit.

CONTENTS

Imagine, to Have Tasted of All the Waters!

A story is told of a man journeying from the north of Europe to experience exotic climes foreign to his native land of ice and snow. His wanderlust brought him to Spain, traveling by train across the splendid Sevilla plain, following the arrow of river pointing to the high Sierra Nevada mountains, to the city of Granada, crowned by the most splendid of Moorish palaces, the Alhambra. To the Moors, who swept out of the parched, windblown lands of North Africa centuries before, the Alhambra was a temple erected to celebrate heaven on earth. Heaven was an endless supply of water, water roaring and gushing, clear, deep and pure, fountainhead of all creation, wellspring of life. Through their fabulous Alhambra the Moors guided sacred water in elegant riverbanks of tiles and glistening marble, spraying through fountains, cascading down intricate carved walls into an endless procession of gardens.

Nowhere in the Alhambra was one ever beyond earshot of water, trickling, gurgling, gushing, always the sound of heaven on earth. The Alhambra was an altar designed to embrace the flow of earth in a thousand forms; as the Spanish poet said, hell is to be blind in Granada. Being a robust spirit, our traveler set off into the mountains above Granada, following the strong river plunging from towering mountain peaks, coming eventually to a town of many footbridges and whitewashed houses nestled at river's edge. A Spaniard came across one of the bridges, a man with a stern eye and stout stick, herding goats before him. The Spaniard spied our traveler and stopped, wondering at this queerly dressed interloper. "Good sir," our traveler spoke out, "I am but a stranger in your bountiful land, and I must congratulate you. Never in my years of wander have I beheld such a landscape to delight the eye and dazzle the soul."

The Spaniard eyed the traveler quizzically, "You have journeyed far?"

"Far, far, far, my friend, all around this sweet earth, to every country that be."

The Spaniard stroked his beard with great deliberation, an odd grin beginning to tug up the corners of his mouth as a look of genuine humility came into his face. He embraced our traveler, tears of joy springing from his eyes. "Imagine," the Spaniard cried, "traveling to all countries that be, to be so fortunate to have tasted of all the *waters*!"

—THOMAS SANCHEZ

Preface

Water is the beverage of the decade. This age-old industry has been revolutionized with millions of new consumers and billions of new dollars. All indicators suggest that it will continue to flow with increasing popularity in the year 2000.

In the late 1970s the stage was set in the United States for bottled water's rebirth. Americans were becoming increasingly aware of fitness and health. T-bone steaks, martinis and midriff bulge were out. The baby-boomers were coming of age and wanted to hold onto their youth as long as possible. Lighter fare in food, in wine, in beverages were part of the solution, and nothing could be lighter than water.

Concurrently, in part propelling the transformation, water became a hot political issue. Tap water in the United States was under fire; the long-accepted belief that American water was safe to drink was challenged by the Environmental Protection Agency, consumer groups and millions of dissatisfied tap-water users. Not only did tap water taste, smell and sometimes look bad, people were getting sick and dying because of contamination.

As bottled water continued to grow through the late 1980s by 15 percent per year, other beverages, spirits for one, were on the decline. Taste buds that had been celebrating bourbon, scotch and brandy began preferring water. Connoisseurs emerged, able to distinguish the range of flavor and styles of this seemingly prosaic liquid.

Water-tasting was nothing new, but suddenly these professionals were in the limelight. Jannette Best, New York City's official water-taster for the municipal supply system, received national recognition because of her ability to distinguish by taste Central Park reservoir water from water dipped out of Bronx's Jerome reservoir or Staten Island's Silver Lake, or Hillview reservoir in Yonkers. Water-tasting supplied trendy stories for national publications and television talk shows. And with every sip the national water awareness grew.

How Do You Describe Water?

Mineral water, spring water, well water, municipal water, carbonated water and even flavored water can now be described and discussed with the kind of enthusiasm and expertise once reserved for wine aficionados. Next time you open a bottle of your favorite H_2O, ask yourself, "Does this water have a slow, medium bead, a random bead, or a profuse, fast bead. Is it crisp? Bubbly? Slippery on the palate? Alkaline or slightly acidic on the tongue? Does it contain a hint of sweetness? Is it sharp and spritzy? Well-balanced? Refreshing? Can it claim a full, elegant mousse, a sharp spritz, a cheerful entry followed by a clean refreshing finish? Of course you will reject brands that have a chemical aftertaste, a metallic finish, sulphurous or chalky aromas, an artificial effect, a taint of cloudiness, or those that are poor thirst-quenchers.

With 500 brands of noncarbonated bottled water and 130 brands of carbonated bottled water to choose from, each with a personality of its own, the language of water-tasting is still growing.

The bottled waters presented in this book represent a selection of high quality and diverse styles. Some are bottled by corporate giants and others are mom-and-pop operations. They are all linked by the quest to provide the cleanest, purest water available; a healthy alternative to our fattening soft drinks, intoxicating spirits and deteriorating municipal water.

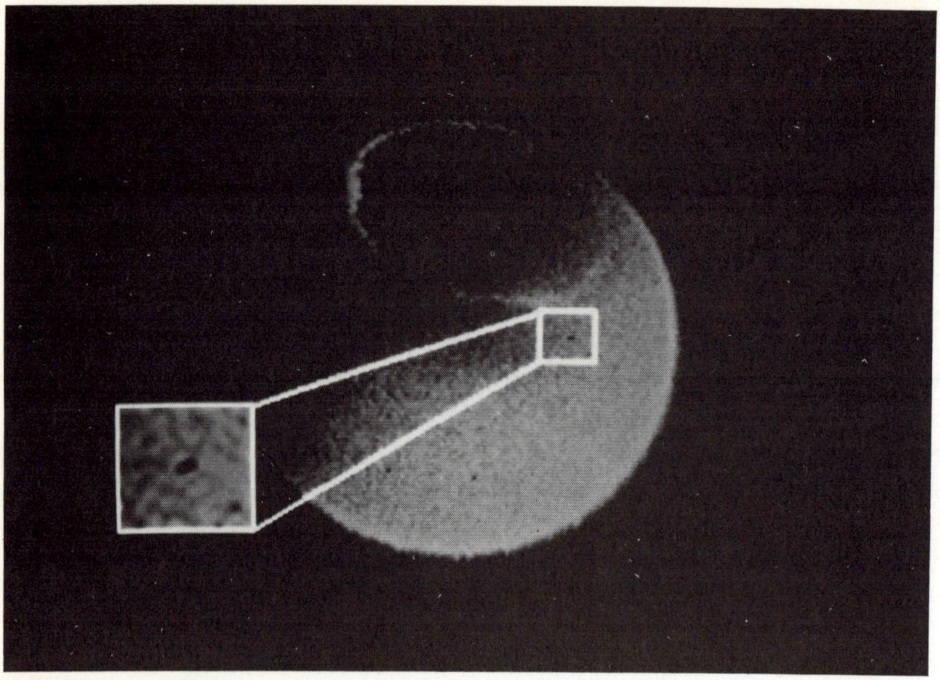

A new theory suggests that water may come to earth from outer space. Satellite photo shows auroral ring in the northern hemisphere. Section in square shows close-up of one of many "holes" thought to be "clouds" of water vapor from small comets that have been disrupted at altitudes just above the earth's atmosphere.

Photo courtesy of L. A. Frank and J. B. Sigwarth, University of Iowa.

H + O_2 = Life

The traditional concept of the origin of water is being shattered by new, revolutionary, scientific findings. H_2O may have arrived as an alien space traveler, according to a recent theory coming out of the University of Iowa. Dr. Louis A. Frank and his team of fellow physicists have discovered evidence that tiny, water-bearing comets, approximately 30 feet in diameter, are colliding with the earth's upper atmosphere, perhaps as many as 20 per minute. Each of these space travelers is believed to carry a payload of about 100 tons of water.

According to the Frank theory, as reported in the May 1986, *U.S. Water News*, the level of earth's oceans may be affected by this outer-space water-delivery system. Hitherto undetected, the annual deposit of water theoretically accounts for all the water on the earth's surface over the stretch of geological time.

A Maverick Hypothesis

Still a maverick hypothesis arising from the space program's polar-orbiting satellite, Dynamics Explorer I, the Frank theory has already found favor with Dr. James A. Van Allen (who discovered the Van Allen radiation belts), and Alexander J. Dessler, editor of *GeoPhysics*.

The Frank theory is in startling contrast to the currently accepted hypothesis that earth's water is born deep within the earth, in its boiling magma — which contains compounds of oxygen and hydrogen, and is released as steam, along with the other gases violently vented during volcanic eruptions.

Whatever its origin, the world's 326 trillion gallons of water exists in forms that change continually, evaporating, then falling to earth as rain or snow, where it remains on the surface or sinks into underground aquifers. Only one percent of earth's water is fresh and in liquid form. The remaining 99 percent is either unusable brine, ice or vapor.

Precipitation waters our world daily. The United States receives 4.2 trillion gallons of it each day, but we lose a little more than half that amount daily through evaporation. And we lose 1,200 billion gallons daily to Mexico or Canada or to the Atlantic and Pacific oceans.

Our surface water flows down two million miles of streams and rivers and dances in 30 million acres of lakes and reservoirs. Fifty times the amount of our surface water is hidden in underground aquifers.

Age-Old Water

Some of our aquifers contain water that has long been out of circulation. "Forty million years old" was the verdict on samples of groundwater taken from 500 to 1,200 feet beneath the Louisiana Gulf coast, dated at the University of Arizona through the use of radioactive iodine 129.

While water's age interests, purity is what affects your health. And you require more water than you might guess; according to one estimate, approximately 1,630,000 gallons of water to feed you for one year. This estimate notes that 12 gallons of water have been used to produce each of your baked potatoes and, if you are a meat-eater, the single steak accompanying that potato represents 1,607 gallons of water!

'Til taught by pain,
Men really know not what good
water's worth.
If you had been in Turkey or
in Spain,
Or with a famished boat's crew
had your berth,
Or in the desert heard the
camel's bell,
You'd wish yourself where
truth is – in a well.

–LORD BYRON

THE STORY OF BOTTLED WATER

1

The Noblest "Element"

"Water is the noblest element."
—Pindar

There is no such thing as plain water. Every drop we drink, unless pharmaceutically purified, contains something more—minerals, gases, organisms, dust particles. Some of these may be harmful, some beneficial, but none should be overlooked.

Water plays a far greater part in our bodies' functioning than most of us realize. Our need for it is second only to our need for air, yet the barrage of tastes, smells, and colors of our daily diet too often overshadows this seemingly prosaic liquid.

This book explains what you will find in your drinking water, how it may affect your health, and what you can do about it. It also reveals the value of mineral waters and provides an encyclopedia of bottled waters available around the world.

In Europe, mineral waters have been recognized for centuries as important to health. This is particularly true in France, where more bottled water is consumed per capita than anywhere in the world.

This is not just a passing fad among the affluent. It is a symptom of a general trend. For the last decade Americans have been growing more and more conscious of their health. During this time we have seen a huge increase in concern about the nutritional content of our daily diet, and the importance of organically grown foods, natural flavorings and balanced diets is now more generally appreciated. It seems inevitable that the value of healthful drinking water will also come to be recognized.

The Nature of Water

Water is a remarkable compound. There is something impressive about the way it combines immense power with complete malleability. The fact that the lower it goes, the greater it becomes (seas are stronger than streams) is a lesson in humility.

Nothing on earth is more serviceable or more powerful than water. It takes whatever comes and reshapes it. The sea slowly reduces rocks to sand, and geologists believe even the highest mountains will eventually be leveled by water, although volcanic eruptions will continue the process of building up the earth's surface.

The Body's Need for Water

Only a quarter of the human body is made up of solids. The rest is water—roughly four to five quarts of it—and we have to keep drinking or become dehydrated, since the body loses water daily through excretion and perspiration. The volume of body water does not change, despite greater or lesser liquid consumption.

Even on a diet, liquid intake should not be curbed. Thirst remains the best guide for your body's requirements. However, insatiable thirst can be a sign of diabetes, and a doctor should be consulted.

Dead Drunk . . . From Water?

In an effort to "purge the devil from his system," a San Antonio, Texas, man OD'd on water, dying of "water intoxication" after consuming 12 glasses of H_2O in rapid order. Speed of intake, more than the amount, was the culprit, according to an investigator from the Bexar County Medical Examiner's Office, creating a chemical imbalance in his bloodstream that resulted in his death.

Japanese travelers pause to admire the power of the Kirifuri waterfall. (Hokusai, c. 1830)

One of the functions of body water is to maintain normal body temperatures. At an air temperature of between 73.4° and 77° F, the body's water loss is about 600 ml per day. Dehydration therefore usually occurs in exceptionally warm places, as the body struggles to cool itself by excessive evaporation. A man can die of thirst in less than a day if he's in a hot enough place, such as Death Valley, California, or the Sahara Desert. In a cold climate dehydration could take up to seven days.

Heat stroke, heat exhaustion, or heat cramps are symptoms of dehydration. When they occur, the victim should be given water and salt, as salt is also lost by perspiration and excretion. A man dying of thirst will experience weakness, lassitude, and—of course—thirst, followed by loss of weight and mental confusion. His cheeks will pale, lips get dry and bluish, and his skin will lose its color and elasticity. His eyeballs will become sunken and the volume of his urine will decrease, while its specific gravity increases. When death is near, breathing will stop. The volume of blood will be maintained at the expense of water within the cells, until death follows.

Water in History and Legend

We take water for granted these days. If we are thirsty all we have to do is go to the kitchen, turn the tap, and the all-provident municipal supply will gush forth in liquid abundance.

It wasn't always that easy. At one time, every drop had to be carried from well, spring, or river. Early man therefore, settled only in areas close to water, which offered transportation and protection from enemies as well as a drinking supply. Later cultures learned how to harness waterpower as they built cities, grew food, and eventually produced hydroelectricity.

Since the need for water was predominant in man's consciousness, it naturally produced

a host of myths, legends, and beliefs that became part of cultures everywhere. These legends diminished as technology hurtled us into the age of science and tap water, but many have survived as testimony to the continuing importance of water in men's lives.

Most cultures had a water god. The Greeks, for instance, worshipped the sea god Poseidon. The Roman version of this god, Neptunus, was not originally a god of the sea, but a god of springs, and so was associated in worship with the nymphs. These were often represented in Roman art as river goddesses wearing shields.

The Babylonians worshipped Ea, the god of the sweet waters under the earth, which, they believed, filled streams, canals and rivers. All fountains and streams were sacred to Ea. Since water was a symbol of life, Ea was also the "Great Physician." It was Ea who befriended Noah, advising him to build a vessel to prepare for the great deluge.

In Egypt, water itself was seen as divine, and it was considered a blessing to be drowned. One of the greatest tributes one could render a god was to be drowned and thus be united with him. The word "drown" originally meant "praise."

Khnemu was one of Egypt's oldest gods, symbolized by a flat-horned ram. At a very early period he was worshipped as god of the Nile and of the annual Nile overflow. He unified in himself the attributes of Rá, Geb and Osiris; his four rams' heads represented the four elements, earth, air, fire and water.

Hāpi was god of both the North and South Nile. The Egyptians paid profound reverence to him as expressed in this hymn: "Thou art the waterer of the fields which Ra hath created, thou givest life unto all animals, thou makest all the land to drink unceasingly as thou descendest on thy way from heaven."

Most cultures also had tales of miraculous mineral springs, which endowed the drinker with youth, health or strength. Perhaps there was some basis in truth for these myths, just as there often proves to be a valid reason for "old wives' remedies."

One such myth relates the discovery of the mineral waters of Bath, in England. The Romans developed the spa there, but long before they arrived the springs were discovered, and the town founded, by an apocryphal figure named Bladud. Bladud was a swineherd who suffered from leprosy, which was cured when he lay in a hogs' wallow with his swine. This wallow proved to be the site of Bath's sulphurus hot springs—and the legend is given credence by the fact that sulphurous waters do, in fact, help leprosy. After founding Bath, Bladud, now a man of stature, was supposed to have fathered no less a personage than King Lear. He then took up the practice of flying around on rocks, and came to an untimely end by crashing into London's Ludgate Hill.

Another myth was told by the Greeks about Hercules. This legend earned him the title of "water-tamer," since it involved taming a destructive spring. The people of the town of Lerna were being terrorized by a nine-headed monster which inhabited the mineral spring called Hydra, on the outskirts of town by the sea. The spring bubbled black water and the monster would suddenly leap forth out of the water, spewing venom and mercilessly killing anyone nearby. One of Hercules' labors was to kill this monster. His efforts were complicated by the fact that as soon as he cut off one head, another would grow in its place. While Hercules was fighting, a giant crab crawled out of the sea and clutched him around the chest. Hercules' charioteer blinded the crab with a burning branch, saving his master.

Hercules finally succeeded in defeating the monster by cauterizing the stump of each neck after severing a head; he then pinned it under a large millstone, thus taming the raging waters to a single channel.

The origin of the Zam Zam spring, in Mecca, is a Muslim legend. It began with Ishmael and Hagar as they were crossing the Arabian desert. After days of walking their water supply ran dry, and Hagar left Ishmael in the shade while she searched for water. She crossed the desert between two mountain peaks seven times, as she believed this would lead her to water. Meanwhile Ishmael began sifting sand through his fingers, when suddenly a natural spring bubbled up out of the sand. The spring, later named Zam Zam, is now in the center of Mecca. During the great Muslim pilgrimage, hajj, which is made during the Muslim month of Dhul-Hijja, a pilgrim approaching Mecca changes his clothes for a seamless garment called an Inram, performs the Towâs (touching the sacred black stone), and drinks from the Zam Zam spring. Ceremonies also include running seven times between the sacred points Es Sâfa and El Marwa.

Water is the very substance of life; early civilizations arose from great river valleys. It is not surprising that water was a focal point in many religions, as here in this Holy Well in the cloister of Megisti Lawra, Athos. (Friedrich Muthmann, *Mutter und Quelle,* Philipp von Zabern, Mainz 1975)

In 1513 Juan Ponce de León, Spanish explorer and governor of Puerto Rico, led his men to the shores of Florida in search of the so-called Fountain of Youth. It was rumored that the local Indians visited this natural spring to be healed and rejuvenated.

Ponce de León never found the spring, and in 1521 he died in Cuba after being mortally wounded by Florida Indians during a colonization expedition.

The following tale about a miraculous spring may not be entirely fabricated. A spring near the river Berkovitza in Bulgaria is supposed to gush blood-red water each time a war is approaching. Four cases have been reported in recent history. The first was in 1885, just before the war in Serbia; the second in 1916, at the height of the Bulgarian World War victories in Macedonia; the third just before the Communist revolution. The last time the waters turned red was in 1935, before World War II.

Water-dowsing is a most intriguing method of locating underground water. Persons known as dowsers walk over the ground, holding out a special forked stick. As they pass over areas where water lies underground, the stick vibrates and in some cases dips toward the ground vibrating violently. Famous dowsers such as Henry Gross (see Kenneth Roberts's book, *Water Unlimited, Henry Gross and His Dowsing Rod*) were hired by communities and individuals to locate, through dowsing, proper sites for drilling wells.

The existence of a secret well at 40 Wall Street in New York is a modern myth, originating with the inception of the Bank of Manhattan, which once did business at that address. The bank drew up a charter in 1799, when it was originally a water company, decreeing that it would develop and supply a water system needed by the city. Aaron Burr, the brilliant lawyer and politician, was the originator of this "water clause," which was really a loophole to

Oceanus appears on two different versions of the corporate symbol of the Manhattan Company. The more elaborate design was used from about 1920 until 1932. In 1933 the annual report was printed with the simpler design, which was closely patterned on the original seal of 1799. (Courtesy Chase Manhattan Archives.)

allow the bank to operate under the terms of that time.

True to its tradition, the Bank continued to use as its corporate seal Oceanus, the sea god, rising from the ground with water pouring from his urn. Once in a while a visitor to the bank tested the water clause by asking for a glass of water; the bank invariably obliged.

Water has often been used by primitive cultures as a rain charm. The Kaitish tribe of central Australia believe pouring water over stones from which a rainbow arose will produce rain. The chief ends the ceremony by pouring water over himself.

In Indonesia, the Torajas of central Celebes have a unique custom for making rain. During droughts they pour water over the grave of a chief. Following that, they hang a bamboo shoot filled with water over the grave. The bamboo is pierced at the bottom so water will drip from it slowly and continually. The bamboo is refilled with water until the rain begins.

Even in certain parts of the United States the practice of pouring water on the last corn cut at harvest—to produce rain for next year's crop—is still carried out.

Water has also been used as a symbol for spiritual cleansing in purification ceremonies. Aztec midwives washed infants with this prayer: "May this water purify and whiten thy heart: may it wash away all that is evil." In India the Brahman rainmaker had to bring himself into union with water three times a day, as well as on various special occasions, to make himself an ally of the water powers.

Water as a Valuable Commodity

In desert countries, water is a supremely valuable resource. Saudi Arabia and other Middle Eastern states have traded it, hoarded it, and even battled over it. As a result of the Saudi oil boom, billions of oil dollars have been spent on the development of an Arabian water supply. The Sand Saline Water Conversion Corporation has run a 15 billion dollar desalination program to meet the country's needs. Saudi Arabia also purchased several of Aqua Chem's distillation plants, to make the largest single distilling plant in the world.

Then, in 1977, Prince Faisal and a French engineer, Dr. Georges Moujin, formed Iceberg Transport International, Ltd. Research began in France and Saudi Arabia to develop a means to transport icebergs from the Antarctic to the Indian Ocean or Red Sea.

Glacial icebergs are enormous potential suppliers of fresh water, some measuring 60 by 200 miles. However, their transportation to Saudi Arabia presents some interesting problems. What, for instance would be the environmental impact? Would it affect the world's water table? And how would the mechanics be managed: what kinds of cables and ships should be used, and should the iceberg be pushed at a point where it is under water to be nearer its center of gravity? Then of course there are the legalities to worry about. Which countries should have legal control of icebergs?

Cloud seeding is another technique for producing water in dry areas. In 1946 the first cloud seeding flight occurred. Vincent J. Schaefer and Irving Langmuir dropped small pellets of dry ice on a supercooled stratus cloud from a small airplane. In minutes they saw the cloud change into an ice-crystal cloud. Small quantities of snow reached the ground.

Since then, silver iodine has become an alternative seeding agent because the crystals are small enough to move with the air and thus have a greater chance of hitting a productive cloud.

In 1947 the Rocking F Ranch in Nevada laid what is believed to be the first claim for water still in the clouds. To supply the 12,000 acre ranch with ample water, the manager planned to seed clouds that passed over it, but before doing so he wanted to secure rights to the water to keep others from laying claim to it.

Because of its scarcity, water was once used as a currency. This happened in central Australia, in the Goldfield District, where workmen took payment in water—eighty gallons for a carpenter for a day's work.

In 1908 at Ocean Grove, New Jersey, the drinking fountain that stood in the railroad station was removed, because when the water was free it was wasted and rotted the floorboards. So tap water was sold to waiting passengers for a penny a glass.

In 1932 the *Evening Post* uncovered a scandal in Washington. Apparently the Senate had been submitting astronomical bills for bottled drinking water. Since the Senate was in session only four months of that year, and the average cost of each senator's water was $35, this caused quite an uproar among the opposition—especially since almost a thousand dollars of the money had been spent on Apollinaris, a German bottled water. Why, they inquired, did the senators have to have imported European waters instead of good, reputable American ones like Saratoga Springs?

The Art of Water Tasting

The wealth of variety in bottled waters and the fact that each spring, each well, each aquifer has subtle taste and scent characteristics of its own, has stimulated great panache in water-tasting in the last decade.

In Florida, a society of dedicated fine-water drinkers, *Les Amis de L'Eau,* pooled resources to sponsor the world's first water-tasting events. Many contests feature bottled waters only, but I was summoned to San Francisco to judge one of the early blind-tasting tap water contests sponsored by the *San Francisco Chronicle.* Water-tasting contests spilled into water-tasting parties, most of them elegant little private affairs, but some are great blasts sponsored by sports associations and health advocates. Wine cellars began shrinking to make room for water cellars.

Fine restaurants, among them New York City's 21 Club, delighted with a dining beverage that doesn't dull the palate, have responded with Water Lists. Nirvana Penthouse and Nirvana Club One, celebrated sister Indian restaurants in New York City, feature a completely separate list titled "Waters of the World," with selections originating from nine countries, including Portugal's Pedras Salgadas and Austria's Voslau Spring.

Of course, it was certain to come and it has come: the H_2O-Bar, where midst a dazzle of colored bottles, customers sit on stools savoring the eclectic subtleties of water, nestled in the back of a very trendy clothing boutique. Not surprisingly, Beverly Hills' Rodeo Drive saw one of the first, begun by d'Urban, a Japanese company, and named with unmistakable simplicity, H_2O-Bar. H_2O-Bar features 77 brand names and is going mail order, with a "water-of-the-month" club.

For your own private water bar you may want to follow the lead of Evian's professional water-tasters who sample its products at 20°C. Water-tasting should be done only with glasses that are well washed and free from cleaning residue. Store your premium waters out of the sunlight, which can cause bacterial growth.

Your "cellar" ideally should be cool, dry and dark. Waters stored in damp areas, especially if they are bottled in plastic, pick up odors. Never store water near chemicals or petroleum products. Stored in warmth and sunlight, water remains potable only for three to six months, but properly stored — in a cool, dry area — it remains healthy, vigorous and taste-tantalizing for five years and more.

A water cellar should provide two important functions. First, it should be a source of emergency water. For that purpose, several five-gallon bottles of water should be kept in your cellar and rotated with each delivery. Only slightly less important, a good water cellar should provide a bottled water for every occasion — sparkling water as a cocktail, a light still water for the meal, and a strong bicarbonate for after dinner.

In the Encyclopedia of Bottled Waters you will find an exciting variety of waters to investigate. Discover those that are available in your hometown, and have a taste test. You will be amazed at the range of tastes, bottle sizes and costs.

A vocabulary to describe different bottled waters may some day be as familiar as the vocabulary of wine tasting: "Full bodied, bracing, soft and salty . . ."

Les Amis de L'Eau enjoy a 1977 water tasting. Left to right are Robert B. Levin, John E. Shuey and Jacques Turner, Honorary French Consul.

2

Pollution and Purity

"Water is a perilous beverage for anyone."
— Charles Mercer, M.D.
from an article in a 1919 edition of *The Practitioner*

Since man first walked this planet he has also polluted it. And he continues to do so. The caveman learned that moving water could help clean his body and carry away refuse. He, like man of today, gave little thought to those who were downstream, drinking that very same water. As civilizations grew and cities emerged, the problem grew worse. Egypt's Nile River, for example, has for millenia carried both disease and death to the thirsty millions who use its waters. Here, traditions, superstitions and religious beliefs support disaster, and unsuspecting residents often draw water from filthy canals, because they believe that the deity would never provide anything that was bad for them, experience to the contrary.

In America, with our victory over cholera and typhus, we believed that our tap water was safe . . . until recently. An article in the *Los Angeles Times* reported that State Health Director Kenneth W. Kizer, in a comprehensive departmental study, found 40 percent of all the contaminated wells in the state are in the greater Los Angeles area. In that same study 2,800 large water systems were tested for the presence of 100 chemicals, and 19 percent had detectable levels of contaminants. With news like this, it is not surprising that the country is becoming increasingly alarmed about the purity of its water.

As far back as 2000 B.C. a Sanskrit manuscript advised that, for safety's sake, drinking water should be kept in copper vessels, exposed to sunlight and filtered through charcoal. More than 1,500 years later, the Father of Medicine, Hippocrates, emphasized the relationship of water to health, recommending the boiling and straining of rainwater. In the 1890s, a project of the Massachusetts Board of Health discovered that slow sand-filtration reduced the death rate from typhoid by 79 percent.

Safety Still a Problem

Almost a hundred years later, the safety of our water is still a serious problem. During the last two decades some Americans have sickened and some have died, from drinking their home tap water. The danger stems from three sources — chemical contamination of groundwater by industry and agriculture, biological contamination of groundwater by humans and animals, and contamination, both chemical and biological, in the delivery system, from source to faucet.

In the 1700s water was considered drinkable if it dissolved soap without forming lumps. Germs in tap water weren't discovered until the late 19th century, when water bottlers in some U.S. cities began delivering spring water by horse and carriage. This deliveryman represented the Crystal Rock Company of Stamford, Connecticut.

Federal and local governments are trying to meet the challenge to save our drinking water, but the public response is clogged with politics and controversy. How safe is safe? What level of toxins and carcinogens in drinking water can be tolerated? What are the effects of long-term chronic consumption of these chemicals? Who will foot the bill to clean up our water?

The commercial interests that produce contaminants often take the position that the diluted pollutants entering our homes via tap water are too minute to cause health problems. Not so, conservationists have warned: even when no immediate effects from drinking polluted water are discernable, over the years, they say, body tissues are being damaged. Events seem to be proving them right. *U.S. Water News*, in February 1987, acknowledged the existence of "... conclusive epidemiological or toxicological evidence ... to prove that certain toxic chemicals do in fact cause specific health effects when ingested in low dosages over long periods of time." Are we at risk when we drink tap water?

Infants and children are most vulnerable to toxic chemicals because the human body's ability to detoxify develops slowly after birth. Nevertheless, most of the testing to fix safety standards was done on adult animals. Are our children at risk when we give them tap water?

The safety of our drinking water casts a shadow in the shape of a dollar sign. Will we pay the price for safe water? Both industry and agriculture, with some exceptions, seem unwilling to undertake expensive preventive measures except when it is required by law. Taxpayers, already burdened, have shown tendencies to heed information campaigns minimizing the need for rehabilitation of their water systems until hit by a frightening local "epidemic" of drinking-water-related disease. Even then the common contention that the disease is merely coincidental with the presence of dump sites over groundwater sources or waste flowing into waterways slows down community response to the problem. Two years, five years of neighborhoods drinking water identified as contaminated while the experts and politicans argue has become part of the history of the last two decades.

Tap water that is safe for infants, safe for the elderly, safe enough so it does not contribute to cancer and tissue destruction over a period of 20 years will be expensive to obtain and expensive to maintain. Robert Burns, sanitary engineer for public health in Sacramento, California, points out, "... Contamination levels of groundwater can vary daily. Though we test groundwater today, it may change tomorrow." A 1986 Harris survey showed 86 percent of Americans are concerned about their drinking water. But testing is expensive; how often should it be done?

Recent events indicate: more often than at present.

Fifty thousand gallons of industrial waste are dumped in the United States every day, according to the U.S. Environmental Protection Agency (EPA); more than one-third in waste sites that are neither licensed nor controlled. Of the *known* unlicensed waste sites, 7,800 are located "directly on top of groundwater sources" with no barrier between waste and ground water. The number of hazardous waste sites added to the EPA's "Superfund" list for federal cleanup reached 702 in 1986, with New Jersey's 90 sites leading the nation.

Slow to Note Health Risks

The EPA admitted, as a result of a hearing before a Senate subcommittee in 1987, that the agency had been too slow in recognizing the public health risks of the Texas Eastern Gas Pipeline Company dumping of PCB (polychlorinated biphenyl) compounds in pits at 68 sites, resulting in contamination that posed immediate danger to the drinking water of local residences in five sites scattered from New Jersey to Kentucky.

Thousands of taps were turned off from Pittsburgh to Wheeling when an Ashland Oil tank collapsed and spilled 3.8 million gallons of oil into the Monongahela River in January 1988. Hundreds of families were evacuated, dozens of factories shut down and schools closed while the Monongahela and Ohio Rivers carried what would later be known as the nation's worst inland oil spill. Emergency water was trucked to collection tanks and the National Guard set up storage facilities for bottled water. Everyone pitched in to get water to the stricken towns; even Funeral Director Tom Somma, of Pittsburgh, used his hearse to deliver bottled water to thirsty townspeople.

One in five Americans drink tap water containing dangerous levels of lead, according to a study released by the EPA in 1987. The study noted slightly lowered intelligence — up to five IQ points — among 143,500 children drinking lead-contaminated water every year, creating learning disabilities in children that will reduce their future wages. In addition to the damage it does to the nervous system, among adults, lead toxicity resulted in 118,400 cases of hypertension, 75 strokes and 370 heart attacks and higher risk of pregnancy complications among 622,000 women of child-bearing age. According to the EPA, lead in drinking water costs more than $1 billion a year in medical care, plumbing repairs and specialized education.

Acting on this recent evidence that long-term consumption of minute amounts of lead is more dangerous than the medical profession had realized, the EPA proposed that the standard of acceptable amounts of lead in drinking water be reduced from 50 parts per billion (ppb) to 20 ppb. Pressure is being exerted on the state level to legislate even lower acceptable levels of lead; California has already reduced its standard to 5 ppb.

Agricultural wastes are becoming an ever-increasing menace to our tap water. Twenty-three states have detected pesticides in public groundwater supplies, according to a 1986 report by the EPA. Since most states have not even begun coordinated monitoring of groundwater supplies for pesticides and their residual compounds, their presence in public and private drinking supplies goes largely undetected. Where the testing takes place, the results are often alarming, as Des Moines, Iowa, discovered in 1987, when tests confirmed citizens' suspicions that pesticides were in municipal drinking water; the California State Department of Health estimates that 20 percent of California's wells are contaminated either by pesticides or industrial wastes.

A Lack of Knowledge

The extent of our danger is masked by lack of knowledge. Some new research reveals the possibilities that long-term consumption of contaminated water can have undesirable results that are more subtle than out-and-out illness. Biologists from the University of West Florida and Stanford University in Birmingham, Alabama, have been studying the "masculinization" of female mosquito fish in Eleven Mile Creek, downstream from paper mills. The female fish even developed a male appendage and began trying to fertilize other females. The two scientists theorize that the woodpulp waste may contain a substance produced by some types of trees that is chemically similar to the male hormone testosterone. No one yet knows what effect minute amounts of that still-to-be-identified substance could have on the development of infant girls or its effect on women, when consumed over long periods of time.

Protecting groundwater, while the largest part of the battle, does not do away with danger to water from its delivery system. Old-fashioned pipes, using lead solder, probably contribute the greatest amount of lead contamination through the water-delivery system. Testing lead levels in the bloodstream is not a usual part of a medical checkup: its symptoms are frequently misdiagnosed. Tap water in all old housing, unless the plumbing is known to have been replaced by newer non-lead piping, should be tested.

The prestige of the Library of Congress failed to protect it from delivery system contamination. After employees began reporting stomach cramps, nausea or diarrhea, the water was analyzed and found to contain excessive levels of lead and copper, believed to have been stirred up from sediment in the pipes when the wing's plumbing had been flushed out with a high pressure hose. Officials had to put up signs warning, "DO NOT DRINK THIS WATER," on drinking fountains in the new Madison wing for almost a month until a new filter could be installed.

At the beginning of the century, waterborne cholera epidemics killed people by the hundreds in frequent local epidemics. Chlorine treatment stopped cholera, as well as typhoid and other diseases of bacteria-contaminated drinking water. Nevertheless, danger from disease organisms in drinking water is not completely behind us. The EPA reports 407 outbreaks of waterborne disease in a recent 12-year period, causing 101,243 recorded illnesses and at least 23 deaths. Experts suspect that ten times that number of victims went unreported.

A New Menace

Chlorine, in the levels used in routine disinfection of municipal drinking water, does not destroy 100 or so viruses identified in drinking water, nor has it proved effective against the latest menace to come out of our taps, *Giardia lambia*, a parasitic protozoan, causing diarrhea-producing infection.

Giardiasis is a difficult disease to diagnose. Its cysts often elude microscopic detection, many doctors are not familiar with it, and in early stages its symptoms — chronic diarrhea, abdominal cramps, fatigue, nausea and intestinal gas — mimic intestinal flu. Often it is not until its chronic diarrhea results in severe weight loss, dehydration and, in some cases, mineral imbalances in the blood stream stressing the heart, that the disease is apt to be finally identified and properly treated.

The Giardia's invasion of our water system is difficult to track because the disease has not yet been listed among those the government requires doctors to report. A giardia infestation in local water can hide behind a mistaken diagnosis of flu. Twenty thousand cases of giardiasis were recorded between 1972 and 1986. But Dr. Richard O. Vogt, Vermont state epidemiologist, warned in 1986, "When other states begin such monitoring programs, they will probably discover what we have discovered here, that the incidence of the affliction (giardiasis) is much higher than expected." After six years of improved diagnosis and reporting, Vermont officials discovered that giardiasis closely rivals gonorrhea as the state's most infectious disease.

Giardiasis poses a danger to camping and hiking enthusiasts. Only the boiling of lake and stream water — one minute at low altitudes to five minutes at high altitudes, or the use of specialized filtration devices specific for giardia cysts, provides reliable protection.

Contaminated lakes and streams often go unreported because people frequently do not become ill for one or two weeks after infection, and often fail to connect their illness to water they drank on vacation.

Once known as "the backpacker's disease," giardiasis outbreaks are occurring in urban areas throughout the country and can be expected to continue until municipal water supply systems are improved, according to federal health officials. Ingested, the heat of the stomach activates the giardia cyst which immediately begins feeding and reproducing, causing sickness and creating more cysts.

Day-care centers that accept infants may be in danger of giardiasis transmission through contact with contaminated diapers.

The infection can last for weeks, or even months, but is rarely fatal. It is successfully treated by Flagyl, an anti-parasitic drug taken orally. Some patients claim the treatment is almost as bad as the disease.

Prevention, the best option, will be expensive. Chlorine levels used in routine disinfection of municipal drinking water are not effective against giardia cysts, according to epidemiologists at the Center for Disease Control in Atlanta, who noted, "The long-term solution to waterborne giardiasis requires improvements in and widespread use of water filtration."

Chlorine Causes Problems

Chlorine treatment of water to counteract biological contamination has other defects. It has been shown to create health problems for people undergoing dialysis. It can combine with other chemicals to produce carcinogenic trihalomethanes. The need to switch to an effective filtration system for purifying water is obvious, but the expense involved and human inertia makes it doubtful that we will see the end of chlorine treatment by the turn of the century.

The news is not all bleak. In May 1987, *U.S. Water News* reported a test being developed at the University of Arizona, using DNA molecules to detect waterborne viruses, that is so simple it may bring new federal standards of water-quality testing.

That year saw a new program promoted by the EPA, using water fleas and minnows to detect water pollution. The agency hopes to see the program extended nationally. In accordance with most state guidelines, industries conduct only a limited number of chemical analyses of wastewater, testing most often for nonhazardous chemicals and compounds. Unfortunately, wastewater does not just disappear; it has been shown to contaminate certain aquifers and wells. The new system can detect both known and unknown health hazards. If the translucent water fleas and round-headed minnows die after living in wastewater for a certain amount of time, the water is not safe for you and me. The EPA is hoping all major dischargers of wastewater will be required to use fleas and minnows to monitor outflow.

Water shortage problems have created pressure in Los Angeles, California, to add reclaimed sewage water to the city's tap water supplies. The effluent would be cheaper than groundwater and, according to one study, just as safe to drink. The study had gone beyond fleas and minnows and monitored humans who had "ingested water from a central Los Angeles aquifer with treated water," yet showed "no increase in birth defects, cancer, or disease." The report failed to note how many people had been drinking the treated sewage, and over how long a time their health had been tested.

According to the September 1987 issue of *Water Market Update*, the year will also be remembered for a landmark decision in which the director of the Idaho Department of Water Resources strongly affirmed the "public trust doctrine," as it relates to protecting water sources. That decision, retaining state rights over water sources, is certain to be quoted as a precedent, nationwide.

Although the subject of much controversy because of its omissions, the Federal Water Pollution Control Act (known as the Clean Water Act), which was passed in 1974, and amended in 1977 and again in 1986, is making a difference. The act expanded the scope of federal responsibility for the safety of drinking water to all community water systems with 15 or more outlets, or 25 or more customers. Its latest amendment set mandatory deadlines for the regulation of key contaminants and required monitoring of unregulated contaminants.

An Expensive Lesson

It also boosted its powers of enforcement, as the officials of Nabisco Brands, Inc., found out in 1987. After pleading guilty to two felony counts of illegal dumping of untreated wastewater into the White River, and making false statements to regulatory agencies, Willard

Spring water can be a valuable source of calcium and other trace minerals such as fluoride, especially for children whose permanent teeth are being formed. Many nutritionists, however, believe there is enough fluoride in other food sources; excessive fluoride intake can cause mottling of teeth. (Mary at the Well, engraving by Meister Matthaus Zasinger, c. 1500. From Friedrich Muthmann, *Mutter und Quelle*, Philipp von Zabern, Mainz 1975).

Kaser, manager of Nabisco's yeast and vinegar plant in Sumner, Washington, was sentenced to a year and one day in prison, and fined $5,000. Additionally, Nabisco was fined $300,000, and forced to establish a $150,000 environmental trust fund, and a $250,000 fund for use by the state's Departments of Fish and Game and Wildlife, to be used primarily in the enhancement of waterways. The company was also placed on a year's probation.

The Resource Conservation and Recovery Act (RCRA) of 1976, with its more stringent 1984 amendments, is attempting to clean up on the dumping of waste on sites that can contaminate groundwater.

The Comprehensive Environmental Response, Compensation, and Liability Act (CERCLA) created the now-famous Superfund to deal with the existing mess of toxic dumpsites. Progress is expected to be slow. The project's 1985 report pointed out, "Costs to Superfund could easily be $100 billion — out of total costs to the nation of several hundred billion dollars, and it could take 50 years to clean up 10,000 sites."

California's Safe Drinking Water and Toxic Enforcement Act of 1986 was an impressive step in controlling pollution through individual state initiative. By no means a shoo-in before it reached the ballot-box, Prop. 65, as it was called in its initiative stage, encountered aggressive, highly-financed resistance. Undaunted, water-drinking Californians linked arms behind the movement, encouraged by some of the most visible members of its entertainment community. The Hollywood Caravan "starring" Linda Evans, Michael J. Fox, Jane Fonda, Goldie Hawn, Chevy Chase, Jack Nicholson, Morgan Fairchild, Rob Lowe and dozens more celebrities toured the length and breadth of the state with all the attention-getting glitter of show-business pizazz, luring thousands of voters to open-air rallies and after-dark meetings.

The California Safe Drinking Water Act prohibits contamination of drinking water sources with chemicals known to cause cancer or reproductive toxicity. It requires that clear and reasonable warning be given to the public by companies prior to any exposure to such chemicals.

To assure enforcement, it reverses the burden of proof from plaintiff to defendant, requiring that chemicals be proven safe by companies dumping their wastes, rather than requiring victims to prove them harmful.

To aid in the detection of harmful chemicals, the act requires the governor to publish an annual list of chemicals known to cause cancer and reproductive toxicity.

Citizens Can Act

Under the act, any citizen may bring legal action to enforce its regulations, and he will be entitled to 25 percent of any fines levied. Offenders will discover that the act has significantly increased fines and jail terms for violations.

Unfortunately, the act does exempt public agencies, small businesses, the military, public water companies, and municipal landfills from compliance. Further, Governor Deukmejian's first list of harmful chemicals published under the act was challenged for containing only 25 compounds. Then, in the summer of 1987, the Natural Resources Defense Council, the Environmental Defense Fund and the AFL-CIO won a court order requiring California to expand the list of chemicals "known to the state" to cause cancer and birth defects to 230.

It is a small beginning. But it is a beginning that "is going to make a massive impact on the country," predicted Dr. Samuel

Epstein, author of *The Politics of Cancer*. It demonstrates that citizen pressure and citizen outrage can create changes to increase the safety of their tap water.

But until those changes take effect, the potential danger of drinking tap water exists. Man will continue to pollute water when he can get away with it. Physical illness, from typhoid fever and infectious hepatitis to tularemia and cancer, will continue to be traced back to impure water. The world's cache of untainted water will continue to decrease, and science will continue to uncover startling new problems with the water we thought was safe. Fear has always been an emotion that bottled water manufacturers have avoided exploiting, but every bottler knows, deep inside, that pure bottled water becomes priceless when a thirsty world is threatened with pollution.

3

Mineral Waters

"Here are your waters and
your watering place.
Drink and be whole again
beyond confusion."
—Robert Frost, *Directives*

A mineral spring is the site of immense activity. By the time water gushes from the ground it has undergone a long and complex process.

With artesian waters, this process begins when ordinary rainwater falls on porous, sandy soil which, deep under the earth, becomes sandwiched tightly between two impervious rock strata. There the porous, water-bearing rock forms an underground pool impregnating the water with minerals in the process. Pressure from the layers of solid rock causes the water level to rise to a point as high as the highest level of the water-bearing strata. When the water is tapped, it may shoot into the air a hundred feet or more.

These waters are named after Artesium (the Latinate version of the ancient French province Artois), where the first artesian well was bored in 1126. Before electric pumps were invented this natural overflow was a very useful method of supplying water in areas where artesian systems were available, such as the Dakotas, Kansas, Australia and Argentina.

Thermal mineral waters differ from artesian water in two main ways. For one thing, thermal springs are hotter. Whereas artesian waters are usually within a few degrees of the average local air temperatures, thermal springs are substantially above it; they can range from 60° F to well above boiling point.

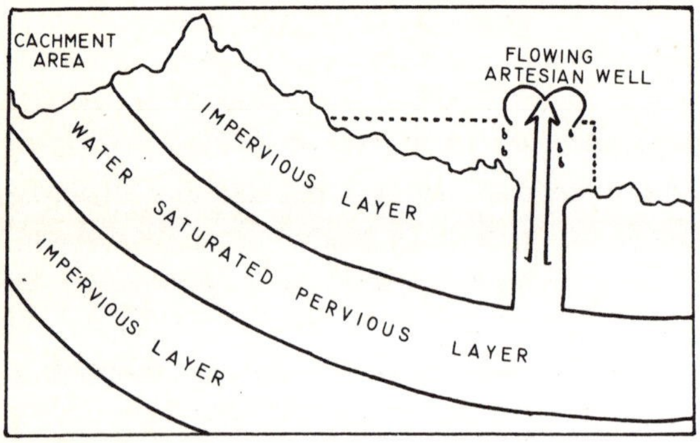

Artesian well. Borings are made into the earth until water is reached. Internal pressure causes the water to flow up like a fountain. Dotted line shows height to which water will rise when released.

And thermal mineral waters are generally much more heavily mineralized (often sulphurous), sometimes to the point where they are undrinkable. They are found near primitive rocks and volcanic regions.

Thermal water occurs when ground water percolates down and is heated by the earth's normal gradient. For every 50-100 feet of depth the temperature rises one degree, which indicates how deep some of these waters must go in order to get as hot as they do. As the waters percolate through the various strata of rock they become impregnated with rare salts and are combined with artesian waters. The waters come into contact with molten rocks, and return to the surface as a hot spring.

Before a hot spring can become a geyser it must have a special underground passageway. The tube does not go straight down to the source of heat, but is twisted and irregular—thus interfering with the natural rise of steam to the surface. The eruption of a geyser is caused by water in a twisted section of the tube being heated to boiling and suddenly turning to steam. The water at the bottom of the geyser is under great pressure from the weight of the water above; pressure raises the boiling point many degrees, so the water is well above the normal boiling point when it finally turns to steam. The escaping steam lifts the column of water from the geyser tube. The overflow lowers the pressure on the water below, which allows more heated water to turn to steam instantly. This steam also pushes upward and more water turns into steam with explosive force. At the earth's surface, a violent eruption takes place. When the geyser tube is filled and again reheated, another eruption takes place.

"Old Faithful" at Yellowstone Park erupts regularly every hour, shooting up to a height of 100-500 feet for about five minutes. Most geyser water contains large amounts of silica and each time the geyser erupts it deposits minerals around the outlet of the tube, which creates a cone.

We know more or less where mineral waters come from, but we are amazingly ignorant about their composition. "Water is the most static substance known to science, yet it has remained as much of a mystery to chemists today as oxygen was 150 years ago." These were the words of Dean James Kermal of New York State University, addressing the American Chemical Society Institute. "Because water is so universal in our own small part of the universe," he went on, "we take it for granted that we know all about it. On this more or less casual assumption we have built up a vast structure of theory when we are completely ignorant of the basis of this assumption. The history of science affords numerous circumstances where whole schools of investigators have gone astray from the neglect of such precautions."

That statement was made in 1927, but the aura of mystery still prevails, despite the progress made by 50 years of scientific investigation. Many ingredients in mineral water are so minute they cannot be detected, and certain interactions between salts and minerals in water are not fully understood. The chemical analyses of bottled waters you will find in section two of this book are accurate as far as they go—but they are still only a general indication of true contents.

Dean Kermal referred to water as a static substance, but in fact this is not entirely accurate. If we could see mineral water at close enough range, we would see its salts seething with activity. Mineral water usually contains a great number of ions (electrically charged particles, or groups of particles). The salts in mineral water are composed of positively and negatively charged ions, which are separate from each other. They are in a state of constant vibration, repelling and attracting each other, and moving from one molecule to another.

To a certain extent ionization approximates a biological system—and in this light one could almost view mineral water, in its constant state of flux, as a living force.

Classification of Mineral Waters

Mineral waters can be divided into five main classes: saline, alkaline, ferruginous, sulphurous and potable. These classifications are not necessarily exclusive, and a given water might have properties which would enable it to fall under more than one heading.

Saline water: These springs can often be recognized by a coating of "salt" around the source. Saline springs are usually found near shale or sandstone, and may have high levels of sodium or magnesium sulphate, or sodium chloride. Saline water high in chloride has been touted as stimulating to the appetite ever since the days of bottled patent medicines; saline water high in sulphates is often used as a purgative, even today.

Alkaline water: Alkalinity in water is characterized by a pH greater than 7. Neutral water has a pH of 7, acid water has a pH less than 7, and alkaline, or basic, water, has a pH between 7 and 14. Typical alkaline table waters vary from 7.2 to 9.5, though mineral waters used medicinally may be more basic. Salts responsible for alkalinity are most usually carbonates, bicarbonates, sulphides and borates of sodium, potassium, calcium and magnesium.

Ferruginous water: Iron bearing, or ferruginous, waters occur in areas of crystalline rocks containing iron compounds. These springs are easily identified by the rusty color of the water, which sometimes stains the soil around the source. Some springs deposit such a quantity of the mineral that it can be used in the manufacture of paint. Ferruginous waters spring from the earth either hot or cold, and are often carbonated. This kind of water poses a problem for bottlers because it discolors both the bottling equipment and the bottle itself. The water may be unappealing to consumers because of discoloration in the bottle.

Sulphurous water: These waters are easily identified by the noticeable rotten egg scent they produce. In some cases the water has a light greenish yellow color derived from complex sulphides in solution. These waters are most commonly used for bathing and not for drinking, and in the past were supposed to cure skin diseases, malaria and leprosy. Taking small doses has been reputed to be helpful in curing arthritis; grandma may have made use of some of these heavily saturated waters when she made up her springtime dose of sulphur and molasses.

Potable water: Potable, or *drinking,* waters, have low mineral content. Water with 500 or more parts per million of dissolved solids are considered mineral waters; those with less are drinking waters. Potable waters are the primary sources for bottled water companies. In some cases bottlers add minerals selectively to give more flavor; or they may add carbon dioxide gas, dissolved as carbonic acid, to lend "personality" to plain drinking water.

Radioactivity in water

At the beginning of the 20th century, attention turned to the benefits thought to be derived from radium in water. At health spas around the world patients were drinking, bathing in and inhaling radioactive water.

Generally the water was found in areas with granite formation. It was initially sold in bottles, but it soon became a common belief that radioactive qualities in water would decay over a period of time if sealed up. It was far more effective to drink the waters at the source.

Health claims went hand in hand with radium treatment. Growth of healthy cells, stimulation of the digestive tract, increasing excretion of uric acid and increasing sexual activity were just a few of the many claims for radium water. Some serious experiments followed the era of radium awareness which turned up some evidence supporting the benefits of radium emanation. A Dr. Halber succeeded in enhancing the development of secondary sexual characteristics of water newts by radium emanation.

A TABULAR VIEW,

Of the comparative Properties of some of the most celebrated **Mineral Waters**, *both in Europe and America ; shewing the number and quantity of Arti Pints, or 115.5 Cubic Inches, of the Water of each Spring : The quantities put down in Grains and Decimal Parts of a Grain, from the latest Aut*

	NAMES OF THE SPRINGS.	Temperature.	Inches carbonic acid Gas.	Inches sulphuretted hydrogen Gas.	Muriate of Soda.	Carbonate of Soda.	Carbonate of Lime.	Carbonate Magnesia.	Carbonate of Iron.	Sulphate of Soda.	Oxyde of Iron.	Sulphate of Lime.	Carbonate and muriate of Magnesia.	Muriate of Lime.	Sulphate of Soda and Magnesia.	Sulphate of Magnesia.	Muriate Magnesia.	Hydro-sulphate Lime.	[illegible]
SARATOGA WATERS.	Congress	50	171.089		235.75	8.25	89.238	1.678	3.084										
	Columbian	50	118.		100.75	11.25	60.5	0.75	3.75										
	Washington	48	123.5		115.75	8.25	63.75	1.25	3.										
	Hamilton	48	142.		134.75	12.25	73.75	3.75	1.5										
	Flat Rock	48	126.25		97.4	5.6	53.5	0.75	3.75										
	High Rock	48	121.36		105.	9.	57.5	0.25	2.25										
	President	51	157.132		67.5	9.5	53.75	1.25	3.										
	Red Spring	48	75.25		58.76	12.24	48.5	1.5	3.										
	Ellis' Spring	47	112.25		94.	5.	55.25		3.75										
	Taylor's Washington	48	125.5		156.77	13.23	70.	2.75	2.05										
BALLSTON WATERS.	Ballston Spa	50	105.		79.5	4.5	37.25	1.25	3.75										
	Low's Spring	52	110.		71.	5.	32.25	0.75	3.										
	Sulphur	52	72.	3.5	32.	3.	15.		2.										
	New Spring	50	159.5		79.25	6.5	30.77	4.25	3.75										
	Saratoga Sulphur	50	21.75	5.5	11.8	0.7	16.55		0.95										
CHELTENHAM, ENGLAND.	Saline Spa	55	15.18		2.5				2.5			20.	12.5		240.				
	Chalybeate Spa	53	12.		1.01				0.68			0.305		1.275					
	Montpelier Wells, No. 1	53	8.5		109.87				3.575	40.005		42.505		18.		49.125	20.		
	No. 2	52	3.95	5.5	91.625		9.	2.875		26.5		33.25		12.08		24.06	14.5	16.375	
	Spa, in Germany—Pouhon Spring		51.16		.688	5.88	5.88	17.84	2.25										
	Germany—Pyrmont		104.		6.24		17.04	40.12	2.24			34.32				22.28			
	Harrogate, England		4.	9.5	307.75		9.25	2.75						6.5		5.25	45.5		
	Aix-la-Chapelle, or, Aken Water	143	some.	much.	20.	48.	19.												
	Bristol Hot Well, England	74	15.		2.		6.75			5.625		5.870					3.025		
	Bath—King's Bath, England	116	4.		26.		11.25		trace.	13.		15.75							
	Bohemia—Carlsbad	165	20.		22.12	47.04	16.00		trace.	166.04									
	Germany—Saltzer Water		70.		70.	16.	12.	20.											

A comparative analysis of various springs in the United States and Europe. This table appeared in the third, revised edition of R.L. Allen's *Historical, Chemical and Therapeutical Analysis of the Principal Mineral Foundations at Saratoga Springs Together with General Directions for Their Use.* (B. Huling, Saratoga Springs, 1853.)

A method of testing water for hardness without having an analysis is to make ice cubes from it. Cloudy cubes indicate soft water, high in sodium. Clear cubes result from mineral free water. A white spot in the center of the cube indicates an aggregation of minerals — the larger the spot, the more minerals.

Other medical authorities of the era, including Engelmann, Von Noorden and Löwenthal, concluded that radium could cure gout, chronic articular rheumatism, gonorrhoeal rheumatism, rheumatoid arthritis, high blood pressure and premature old age.

The *British Medical Journal* of January 25, 1913, reported that through radium emanation treatment of arthritis deformans could be "strikingly benefited by the daily drinking of 250 cc of radium emaratis solution of a strength of 1-2 millicuries per liter. The treatment must, however, be persisted in for quite a long time, and at least six weeks are likely to elapse before any change is noted."

Many people became fearful of radiation after the successful testing of an atomic bomb in the New Mexico desert on July 16, 1945. The bomb introduced into the atmosphere radium 226, radium 228 and strontium 90, which subsequently were found to be carcinogens according to the EPA. Federal regulations have established maximum allowable concentrations in bottled water. (See Appendix A, *Bottled Water and the Law.*)

Carbonation

Carbonated springs are found in regions rich in lava, as well as in areas where calcium carbonate, or limestone, and other carbonate rocks are common. Water bottlers retain natural carbonation by first removing the gas from the water at the source and later re-injecting it into the bottle in the bottling plant. Usually the gas is dissolved under pressure of three atmospheres. Many bottlers artificially supply their water with carbonation simply by adding CO_2 under pressure to the product. In several isolated cases, water is bottled just as it comes from the spring; thus the carbonation level is very low. When dissolved in water, the gas forms carbonic acid which is believed to increase the rate at which fluid is absorbed by the stomach. (This might explain why one is affected so fast by champagne!) Carbonated beverages are noted for relieving indigestion. Carbonation gives bottled water a pleasant sharp taste.

Hard and Soft Water

Mineral water is hard water. This means it contains calcium and magnesium salts, collected as the water passes through the earth's crust. Limestone deposits are a major source of calcium. These two salts are responsible for the build up of tough white scale you may find in your water pipes. They prevent soap from lathering well, and form sticky curds which cling to clothes.

Hard water causes spots on glassware, due to mineral deposits which are left behind as the water evaporates. And hard water reacts with food on dishes, making them harder to clean. Hard water can cause an additional problem in the laundry: iron and manganese in particular can make fabrics stiff and gray. If you live in a hard water area, cleaning problems can be avoided by using extra-hot water, special detergents with softening agents in them, or softened water in the laundry and bathroom.

While hard water is bad for cleaning, as a *drink* research has proven it to be a lifesaver. Hard water, in fact, may prevent your dying of a heart attack.

Twenty years ago, Andrew Shaper wrote, in the *Journal of the American Medical Association,* "It seems difficult to believe that anything as ordinary or as natural as the purified water we drink could be potentially dangerous. But as evidence accumulates it looks more and more likely that there is something other than the drinking water, or closely associated with the drinking water that affects the rate of heart attacks and strokes in this country."

By 1960, that "something" had been traced

to the hardness or softness of the water. Dr. Henry Schroeder, writing in the same journal, reported that he had found areas with hard water, including the Mississippi, Missouri, and Ohio river basins, to have lower death rates than areas near the Atlantic and Pacific seaboards, which had soft water and higher death rates. He concluded, "Some factor, either present in hard water, or entering in soft water, appears to affect death rates from degenerative cardiovascular disease."

Schroeder's factor has now been identified as magnesium. Twenty years ago, very little was known about this mineral, but today deficiencies in it have been found capable of producing heart disturbances that range from premature beats to deadly ventricular fibrillation. Patients suffering from abnormal heart rhythm show marked improvements when given magnesium supplements.

Research in the United States, Britain and other countries has shown that death rates from heart disease are lower in areas using hard water. In 1975, Toronto researchers discovered a marked increase in magnesium concentrations in heart muscles in hard water areas. Two years earlier, researchers had found a high mortality rate from heart attacks in Finland occurring in a clear cut regional pattern. In eastern Finland the death rates were nearly twice those in the southwestern part of the country. In 30,000 soil tests it was found that the soil in southwestern Finland contained an average of 300 miligrams of magnesium per liter. This level declined gradually in an eastward direction, so that in the high mortality areas there were only 100 milligrams of magnesium per liter of soil. The levels of calcium and potassium were also higher in southwestern Finland. "Accordingly," the researchers reported in the *British Medical Journal,* "In Finland there is an apparent inverse correlation between mortality from ischaemic heart disease and the soil content of calcium, and potassium, and in particular, that of magnesium."

Softened water is not only dangerous for heart patients, but also for those with kidney conditions. Water is softened by the addition of sodium, through the cation exchange process (see glossary). The harder the water, the more sodium is added. In 1967 the New York City Health Commissioner sent a letter to 18,000 physicians warning them that softened water has a higher salt content than hard water; patients with heart and kidney ailments might be getting too much salt from softened water. He went on to recommend they buy bottled water or install a system to supply unsoftened water for drinking and cooking.

Tooth decay is also less prevalent in hard water areas, and this may well be attributed to trace minerals other than fluoride.

All in all, it seems that Dr. Arthur Schoch, authority on trace elements in drinking water, was right when he claimed, "People should drink only hard water as it is good for their teeth, bones, blood vessels, and it tastes good as well."

Water hardness is usually measured as follows:

	Grains per gallon	*Parts per million*
Soft	0– 1/2	0– 9
Slightly hard	1/2– 3	10– 60
Medium hard	4– 6	61– 110
Hard	7– 12	111– 200
Very Hard	Above 12	Above 200

Mineral Water and Health Cures

For centuries people from widely differing cultures have believed in the curative properties of mineral waters. Indians in North America have placed great store by mineral springs, and the Romans were responsible for the popularity of warm baths in Europe, where they built a great number of bath houses over thermal springs. For centuries since then the noble and the not so noble have visited steamy hot mineral baths for the recuperative powers they possess.

Bath, a four-letter Anglo-Saxon word, names a town in England famous for that occupation, to which it owes its existence.

The Social Security administration of some European countries will reimburse patients for bathing cures. They believe it reduces pharmaceutical expenses.

In England in 1672 the King's bath (right) and Queen's bath (left) were social gathering places for the royal court. (From a drawing by Thomas Johnson.)

The name that became a town first attached itself to a natural hot springs (heated by a volcanic fault), the only one in all that cold, blustery, foggy island. Understandably, the hot springs of Bath have been well-used since before written history. The Romans delighted in them during their occupation, commemorating their joy in columned bathing pavilions and a temple to Minerva. The spa and town that grew around it flows through British history. The English crowned King Edgar there in 973; two monarchs, Richard I and Elizabeth I, granted Bath a charter; and Charles II used its waters to wash away his notorious dissipations. The consort of James II, Mary of Median, who feared she was barren, after going to Bath "to take the waters" was able to boast that she "was breeding."

But the spa owes its literary fame to the eighteenth century when it became a social cult. Anyone who claimed to be anyone at all must be seen at Bath during "the Season," where there were balls and masques, gambling and promenading in open carriages. Reputations were made and lost in displays of clothesmanship, horseflesh and driving skill.

Pepys recorded being carried to Cross Bath at 4 a.m. where he was both "pumped" and "bucketed" while watching ladies bathing in their canvas covers and men in coat and drawers. Defoe recorded floating trays topped with nosegays for the ladies, snuff for the gentlemen. And Anstey, in 1766, left his impressions in verse:

> T'was a glorious sight to behold the
> fair sex
> All wading with gentlemen up to
> their necks.
> And today many persons of rank
> and condition
> Were boiled by command of an
> able physician.

But today Bath shimmers only in memory of its illustrious past. When I visited the King's Bath a few years ago, my guide warned me not to get near the famous waters, which were at that time closed to visitors. I was told that a young girl had contracted meningitis from the water, resulting in the health department closing down the baths.

Mineral Water Spas

While Bath wallows in reorganization, other spas are beginning to flourish once again. The symbiotic relationship between spas and bottled mineral water cannot be avoided. Today's great spas like Royal Club Evian, Vittel, Contrexeville and Fuiggi all bottle their own waters. Some European spas offer a selection of bottled waters, each with its own therapeutic value. In Europe, and other areas of the world, bottled water has recognized medical value. Those values are determined by the various ministries of health and are printed directly on the label as well as used in advertising.

Evian is recognized as water especially good for infants because of its light mineralization, while Vittel Grand Source is considered especially effective for the digestive system of overindulgent adults. In spas, mineral waters are given to patients on a prescribed program depending upon the individual's needs. Freshly bottled mineral water is combined with treatments that range from underwater massage, hydrotherapy and mineral baths, to futuristic laser therapy. The results are remarkable, from reduced weight and better digestion to increased energy and sense of well-being. The December 1987 issue of *Shape Magazine* established a list of the world's best spas from Lake Geneva to the Santa Catalina Mountains near Tucson, Arizona, the one common denominator of them all being that after a visit the client returns home invigorated in mind and body, more attractive and looking forward to a return visit.

Today there are approximately 2000 spas in the United States alone, and perhaps another 6000 around the world. Every year these spas attract more and more guests; their popularity is once again on the rise. People are seeking health and relaxation in their efforts to cope with the fast-paced modern world. As travel becomes more readily available, and increasingly affordable, a trip to the spa of Baden-Baden, Montecatini, or La Costa becomes a viable alternative to buying a second car or an addition to the home. Banks will actually lend money for a spa visit, and in some cases the trip can even be tax deductible as a medical expense.

Are mineral waters really curative? They certainly contain many elements that have been proven essential to health. (See *Minerals and Trace Elements,* at the end of this chapter.) There is some controversy over whether these inorganic compounds can actually be absorbed by the body. Many American doctors challenge the assumption that either drinking mineral water or bathing in it can result in absorption of water's mineral content.

Japanese, French, Russian and Italian researchers, however, have shown experimentally that the body does absorb certain chemicals and minerals, both through skin contact and by digestion.

In their book *Minerals: Kill or Cure*? (Larchmont Books, New York, 1974), Ruth Adams and Frank Murray report that, "the human body must have minerals and they must be replaced daily." For example, in Europe several years ago athletes began a cross country run declaring they would consume nothing but distilled water. Halfway the runners had to discontinue drinking distilled water; they had lost their mineral reserves including vitally important electrolytes. To continue might have proven fatal.

Some scientists believe water may act as a catalyst and through the domino principle result in restorative reactions. In Dr. Henry A. Schroeder's book, *The Trace Elements and Man,* he states, "It is very likely that cadmium in drinking water and other fluids is more readily absorbed from the intestinal tract than cadmium found in foods."

In the United States, as well as in several other nations, laws prevent spa owners from claiming their waters possess curative properties. A hundred years ago the Mountain Valley Water Company advertised its water as a remedy for many ailments, and boasted many cured patients. The Food and Drug Administration restricted these claims on the grounds that independent researchers could find no convincing evidence that Mountain Valley water could cure disease. Mountain Valley waged a quarter-of-a-million dollar legal battle with the FDA trying to get recognition of facts and testimonies from cured patients that their water really can cure.

There is a tendency in the United States for official organizations to avoid giving credence to anything that smacks of folk medicine. Mineral water health cures unfortunately fit into this category, and we must await further research before the government, and physicians' organizations, will recognize there may be truth in folk medicine after all.

In Europe and, despite the FDA, even in America, thousands do flock to spas to "take the cure": usually bathing at various temperatures. In same cases herbs — often eucalyptus — are added to the baths. Bathing is usually done under medical supervision, since staying in too long can be harmful. Patrons drink the water as well as bathe in it; eight glasses a day is a typical dosage.

Purgative waters have been available for

Due to her failing health, Queen Charlotte of England decided to take the waters at Bath upon recommendation of her physicians. Behind Her Majesty's chair stands the Duke of Clarence who suggests that the flavor of the water could be enhanced with a little brandy! A lady offers the Queen "Old Tom," another phrase for gin. The excitement generated by Charlotte's visit brought this caricature to press by S.W. Fores in 1818.

many years at pharmacies in the United States. One highly esteemed water was called Pluto Water, distinguished by a red devil on the label.

When to Drink Water

Should one drink water before, during, or after a meal? No one seems able to agree on this. Some people believe water should be drunk before the meal, allowing time for the stomach to remove the chill before eating. In 1928 managers of several large restaurants instructed their staff not to serve water until the rest of the meal had been eaten. They claimed drinking water before the meal would chill the stomach and cause gas and indigestion. Or perhaps the clever restaurateurs thought a client would order more food if his stomach was empty. In the same year Dr. Meter Golob claimed drinking water during a meal was unhealthy. He believed water would stretch an already loaded stomach, and went on to assert that water was not absorbed by the stomach unless it was empty. Furthermore, he claimed, water diluted the digestive juices.

I believe you should drink water whenever your body tells you to—whether before, during, or after a meal. The body knows best when it needs more liquid. It might be mentioned, though, that drinking *very* cold water is not advisable, since it may cause indigestion by solidifying fats and oils. It also forces the stomach to heat the chilled water, a wasteful use of body energy.

I feel it is a good idea to drink several glasses of water when you wake up. This helps cleanse the digestive system after a night's rest. I always do this, and have developed the habit of keeping a quart bottle of water at my bedside. When I retire I have a glass, and if I should wake up thirsty in the night (I'm a "midnight drinker"), the water is right there.

Water in Food Preparation

The value of good water in making a successful recipe has long been known to cooks. The English, that nation of tea drinkers, know the quality of water can make or taint a cup of tea.

Several years ago master baker and chef Carl Rosenberg was appointed Chief Baker for the Century Plaza Hotel. One of his first assignments was to see if he could duplicate the famous dinner rolls made at the Olympic Hotel in Seattle. Rosenberg went to Seattle, and on returning to Los Angeles tried out the recipe. The rolls' distinctive flavor and texture were both missing. He rechecked the ingredients, but all were correct and of the highest quality. "Could it be the water?" he wondered. After ordering several gallons of Seattle water, he tried the recipe again. This time the rolls came out exactly right. Since it was not practical to ship Seattle water to Los Angeles, Rosenberg began to bake with purified water. He found that deionized water produced better dough fermentation and enhanced the flavor of his baked goods.

In Miami, Lou Kaplan, owner of the candy store House of Nuts, began using bottled water to make his Lady Fern fudge. It was such a success that he secured 200,000 gallons of Fountain of Youth water for future use. Every box of the nationally distributed candy identifies the bottled water as one of the ingredients. There is even a water cooler in the store with a sign above it inviting customers to try Fountain of Youth water.

In 1965 Star Kist began canning tuna in Arrowhead Puritas Spring Water. The company had conducted tests to determine whether tuna could be canned in spring water instead of oil. The results were satisfactory, and Star Kist now uses about 1,000 bottles of spring water every month.

Bottled Water for Cosmetic Purposes

Cooking, drinking and bathing are not the only uses for mineral water. In France, Evian water has long been used for cosmetic purposes by both men and women. The French spray it on all over to cleanse, cool and beautify. They find it is necessary to replace the moisture that evaporates because of dry air, in the sun, in windy weather or in an overheated room. Evian's Brumisateur is just pure mineral water which is propelled by nitrogen gas. In France it is not uncommon to see a person driving in traffic, while spraying his face with a can of Brumisateur.

A well known Beverly Hills beautician has opened a salon using bottled water for cleansing, because it is so pure. He claims expensive after shave lotions are unnecessary, since refrigerated pure water is one of the best after shave products a man can use.

Minerals and Trace Elements

In the encyclopedia section of this book, you will find water analysis tables listing various chemicals and minerals contained in each water. The following brief description of some of these elements will help you understand their importance in your drinking water.

Only beneficial elements, and those found in many of the bottled waters, are listed.

Calcium

Research indicates that adequate calcium early in life not only builds a strong frame, it keeps the frame strong in old age. Teeth are as dependent on adequate supplies of calcium as are bones.

The small amount of calcium that is not

Evian's Brumisateur mist helps in moisturizing the skin and can also be used as an after shave product or for cooling off in hot weather.

bound up in the skeleton and teeth is extremely important because it stimulates nerve and muscle function and helps control blood clotting. It helps maintain the blood's proper pH balance and activates several enzymes. Lack of calcium in the blood may cause muscle weakness, irritability or increased sensitivity to pain. Low levels of either calcium or magnesium seem to promote lead accumulation in children.

Magnesium

Magnesium is an effective tranquilizer and muscle relaxant. People are discovering that adequate amounts of magnesium eliminate the need for sleeping pills. Magnesium relieves muscle cramps, spasms and tremors. By helping the smooth muscles around arteries to relax, it plays a role in reducing high blood pressure. The possible use of magnesium to prevent migraines and strokes by relaxing constricted blood vessels is being investigated, as is its use in alleviating premenstrual syndrome. Not surprisingly, people who have discovered magnesium report lessened stress, fatigue, irritability and apathy. Magnesium plays a role in preventing both osteoporosis and certain types of arthritis by regulating the hormones that keep calcium from being absorbed into soft tissues, and increases calcium absorption into bones. Magnesium's role in regulating calcium absorption led Guy Abraham, M.D., a research gynecologist and endocrinologist, to recommend a magnesium-rich diet for people with calcium deficiency problems.

Inadequate dietary magnesium may lie behind the craving of chocolate addicts, suggests Nan Kathryn Fuchs, Ph.D., author of *The Nutrition Detective.* Noting that chocolate is the highest in magnesium of all foods, Dr. Fuchs predicts, "Your chocolate cravings will vanish when you have enough magnesium in your diet."

The word *magnesium* is derived from the ancient Greek city of Magnesia, situated near large deposits of magnesium carbonate. One of its earliest known medical uses was as a laxative. An essential part of over 300 enzyme and hormonal functions, magnesium is found in abundance in the human body, where it is used in energy production, muscle strength, protein synthesis and proper nerve function. Symptoms of magnesium deficiency are sleeplessness, muscle cramps, irritability and depression.

Stress, alcohol, sugar and heavy perspiration deplete magnesium supplies in the body. Elemental magnesium is more easily absorbed in liquid form than in solid form. At best only about 30 percent of elemental magnesium is absorbed, while the rest is eliminated. Herbicides, pesticides and fertilization with nitrogen-heavy rather than balanced soil conditioners, work to reduce the magnesium (and other minerals) content of vegetables and cereals, a deficiency passed along to livestock.

Potassium

Dieters be warned, the heart needs potassium to function properly. Potassium, found mainly inside cell walls, helps the body cope with sodium, which is mainly found outside cell walls. Too little potassium permits sodium to break through the cell walls, creating havoc.

Reduced calories in meals can also reduce the usual intake of potassium. People who eat fast food, highly processed food, food that is high in fats and high in salts can also benefit from obtaining potassium in their drinking water.

Potassium works to control the activity of muscles, nervous system and kidneys. It prevents over-acidity in the blood and tissues and promotes secretion of many hormones.

A potassium deficiency can cause muscle

weakness, fatigue, constipation and listlessness. An estimated average daily requirement for potassium is .5 gm.

Sodium and Salt

Sodium — the positive ion of ordinary table salt — helps maintain normal osmotic pressure and water balance in your body. In excessive quantities, however, it has been shown to be a factor causing hardening of the arteries and high blood pressure.

However, Floyd C. Rector, M.D., professor of medicine and physiology, University of California, San Francisco, pointed out, "In recent years, attention has focused on the possible harmful role of dietary sodium, particularly in individuals with heart disease or high blood pressure. What has not been widely recognized, however, is that it is not sodium alone that is potentially harmful in these individuals, but the combination of sodium and chloride. For example, in individuals with high blood pressure, sodium given along with chloride will increase the blood pressure, whereas sodium given along with bicarbonate has no effect on blood pressure." More and more evidence suggests that salt (NaCl), not sodium (Na), is the culprit. Sodium alone, especially in sparkling water with its component of bicarbonates, may be innocent of wrong doing.

Americans are second only to the Japanese, unfortunately, as the world's largest consumers of salt, with an average daily intake between six and fifteen grams. A common estimate for minimal sodium needs is about .5 gm. per day; when performing heavy work in hot climates, 10-15 gm. a day are needed, according to the Food and Nutrition Board of the National Research Council. Over 100 ppm of sodium in drinking water is too high for a baby; over 1,000 ppm is considered too high for anyone.

Zinc

Essential to the formation of RNA and DNA, and for the synthesis of body protein, zinc is a vital element in promoting growth. Delayed sexual maturity, impaired sexual function, spontaneous abortion and enlargement of the prostate gland have been related to low levels of zinc. Essential to immune system function, zinc aids resistance to infection, wound healing and prevention of skin disease.

In areas in the world where soil is especially poor in zinc, such as in the Middle East, dwarfism is not uncommon. Zinc is an important component of connective tissue, both in the skin and the joints. A zinc deficiency may result in stretch-marks in the skin and joint pain. Improved intake of zinc has restored the sense of taste and smell. Zinc deficiencies have also been linked to hair loss and dandruff. At one time, many homes had zinc-coated (galvanized) water pipes, and water passing through these pipes picked up a significant quantity of the mineral. Today, with the increasing use of plastic piping, this healthy addition to drinking water has largely ceased.

Micronutrients: Small Necessities

Chromium functions in metabolism, helping to boost energy and regulate blood sugar levels. It assists in the utilization of vitamin A.

Cobalt forms part of vitamin B_{12}, which is necessary for neurological function and to the manufacture of red blood cells.

Copper helps form hemoglobin in red blood cells. It is needed for wound healing and bone growth. It is a component of elastic muscle fiber. It acts as a catalyst for metabolizing energy on a cellular basis. It functions as an antioxidant. Overdoses can create neurological problems. Deficiencies bring on anemia, low white blood cell count, skeletal disorders, edema, fainting, emotional upheaval.

Fluorine strengthens teeth and bones. In excessive amounts it can cause fluorosis, mottled teeth, bone defects and abdominal pain.

Iodine is necessary for proper thyroid function. It prevents goiter and is used in the structure of many enzymes, including insulin.

Iron is required for cell function. It is a necessary part of hemoglobin, functioning to transport oxygen in the blood. It is essential for the immune system, and assists skin and hair growth. Deficiency produces anemia, fatigue, infection, poor wound healing and constipation.

Manganese is an antioxidant that is found in the brain, glands, bone and cartilage. It is needed to metabolize protein and fat. It activates a number of enzymes and is necessary for brain function. Overdoses produce behavioral problems. Deficiency results in a slowdown of body and brain

---

Essential bulk and trace elements on the earth's crust and in man.*

	Crust (Parts per Million)	Man (Parts per Million)	Daily requirements in food and water.
Bulk or Major Elements			
Calcium	36,300	14,000	1.1
Sodium	28,300	1,600	4.4 (0.2)
Silicon	27,720	260	0.003
Potassium	25,900	2,000	3.3
Magnesium	20,900	290	0.31
Sulphur	5,200	2,300	0.85
Phosphorous	1,180	12,000	1.4
Chlorine	200	1,400	5.1 (0.4)
Trace Elements			
Iron	50,000	60	0.013
Manganese	1,000	0.2	0.003
Fluorine	700	37	0.603
Chromium	200	0.2	0.0005
Zinc	65	33	0.013
Copper	45	1.0	0.005
Cobalt	23	0.02	0.0003
Molybdenum	1	0.1	0.0002
Iodine	0.3	0.2	0.0001
Selenium	0.09	0.2	0.00001
Vanadium	110	0.3	?
Possibly Essential Trace Elements			
Nickel	80	0.1	?
Arsenic	2	0.1	?

*From *Trace Elements and Man,* Dr. Henry A. Schroeder, The Devin-Adair Company, Old Greenwich, Conn., 1973

energy levels, inadequate production of acetylcholine (a neurotransmitter), dull, clouded thinking, unconsciousness.

Molybdenum is used by enzyme systems. It plays a part in regulating growth and preventing dental cavities. Deficiency is thought to be a factor in some male impotency.

Nickel may be necessary for hair and bone growth.

Sulphur is used in the production of energy, and insulin function. It keeps cells chemically balanced, protects cell membranes, and assists in the blood's heparin activity. Deficiency produces irregular menstrual cycles, dry itchy skin, dull hair, badly formed nails and a possible slowdown of the immune system.

Reading an Analysis

You don't have to be a rocket scientist to understand a mineral analysis, but it helps. To help you know which minerals your favorite bottled water contains, wherever possible I have included a mineral analysis of each featured water.

These analyses were provided by the bottler and are required by most countries around the world where water is being sold commercially. In the United States, the Public Health Service monitors bottlers and each state requires a laboratory analysis on a regular basis.

Laboratories charge celestial prices for priority pollutant tests created to detect such undesirables as pesticides, polychlorinated biphenyls (PCBs) and asbestos. It is easy to incur many thousands of dollars on an analysis of the water from one source. If the water is being bottled and sold commercially, it is safe to say that it is safe to drink. Therefore, the question of learning to read an analysis becomes more one of nutrition and gastronomy than of safety.

In a mineral analysis report the water's characteristics are expressed in parts per million (ppm) or milligrams per liter (mg/l) which are equivalent. How much is one part per million? According to William B. Deichman's *Toxicology of Drugs and Chemicals*, one ppm equals:

one inch in 16 miles;
one minute in two years;
one penny in $10,000;
a needle in a ton of hay;
one mouthful of food in a lifetime of eating.

And yet, less than one ppm of hepatitis B can infect us with a painful and lengthy sickness!

Today, laboratory equipment has explored new dimensions of microanalysis and can show parts per billion and even parts per trillion.

Some waters show a pH value. The pH is a chemical symbol expressing both acidity and alkalinity. The pH values run from 0 to 14; 7 indicates a neutral pH. Numbers less than 7 show increasing acidity; numbers over 7, increasing alkalinity.

Nutritionally, certain minerals and trace elements have been proved beneficial to our health. A summary of information about the minerals you find in some bottled waters, and the results of the recent investigations of calcium and magnesium will be given later in the book.

Also worth remembering is the Federal Drug Administration's terminology relating to sodium:

Low sodium = 35-140 mg per 8-ounce serving
Very low sodium = 5-30 mg per 8-ounce serving
Sodium-free = 5 mg per 8-ounce serving

Hard water — i.e., mineral-rich water — has been especially celebrated in research by Dr. Henry P. Schroeder, while at the Dartmouth Trace-Element Center, who reported that people who drink hard water are less likely to die of cardiovascular diseases. Subsequent studies in England, Canada, Sweden and Japan have confirmed that drinking hard water is healthier than drinking soft water or mineral-free water. You know you are safe when you drink bottled water, but by learning to read labels and choosing your favorite brands with care, you can improve your health and vitality.

4

The Bottled Water Business

"They that go down to the sea in ships and do business in great waters."
—Psalms 107.23

Bottled water has become a billion-dollar business. To be precise, in 1986, a $1,400,000,000 business! Sales are increasing vigorously year-by-year: 1986 saw a 20 percent increase; a 15 percent increase was anticipated for 1987. One out of fifteen Americans use bottled water as the *primary* source of drinking water, according to an International Bottled Water Association estimate. What kind of drinking water is America's favorite? The country has divided itself into passionate sparkling-water advocates and noncarbonated water advocates, with both camps growing at approximately the same rate.

The bottled water business has been the arena of major corporate acquisitions in the 1980s. According to *Beverage Marketing*, an industry publication, in 1986, The Perrier Group led the pack with an estimated 40 percent market share. Perrier acquired several large domestic bottled water companies throughout the 1980s in both premium mineral water as well as in the five-gallon-bottle water business. When Beatrice Foods "liquidated" its water holdings in 1987, McKesson moved into second position as a leading bottled water company, followed by Hickley & Schmitt, Sierra Spring, Kentwood and Suntory International.

At its projected growth of 15 percent, investing in water has been making good sense. Outside investment has not only come from The Perrier Group, but also from Japan's Suntory, which purchased Kentwood in 1986 and Beverage Management's Water Division at the end of 1987, as well as from Switzerland's Nestle, which coventured with France's Vittel and began marketing Vittel Bartlett. Water stocks are now traded on the various exchanges, and large beverage companies like Anheuser-Busch, Coca-Cola and Pepsi have their own bottled water divisions.

There are two distinct bottled water markets, each with a myriad of offspring. Largest in volume are the five-gallon home and office cooler type, often called jug water. The five-gallon bottle, which until recently was packaged in glass, but now more commonly in polycarbonate, is unique to the United States. There is currently some test marketing of this size container abroad, but in most European countries two liters is the maximum allowable size. According to the International Bottled Water Association, five-gallon water products are primarily sold as an alternative to tap water.

Many water products packaged in bottles are also available in smaller packages for purchase at grocery stores. Some bottlers use one-gallon plastic containers similar to those used by milk companies. Some companies, such as Arrowhead, offer a two and one-half gallon bottle with a spout which fits in the refrigerator and dispenses water like a minicooler.

Premium still water is a more elegant version of bottled water suitable for hotel, restaurant and bar use. Leaders like Evian, Vittel Grande Source and Mountain Valley have demonstrated the worldwide appeal of the liter or smaller-size bottle for both service industry and grocery sales. Usually, premium water is bottled from a spring, and is available in glass and high-grade plastics.

Either naturally or artificially-carbonated premium bottled water is another important segment of the water industry. The Perrier Group must be credited with establishing the boom in premium carbonated water. In the mid-1970s Perrier began aggressively marketing their mineral water with a certain style and chic that America thirsted for. Combined with a growing health consciousness, Perrier created a natural, noncaloric alternative to cocktails and soft drinks. Other imported mineral waters saw the potential in the United States and began marketing their products more aggressively. Italy's San Pellegrino and Sweden's Ramlösa attacked Perrier's market. Domestic water companies were created or revived and brands like Calistoga, Saratoga, Artesia, Santé, and Crystal Geyser battled Perrier on supermarket shelves across the country.

Even companies that bottled club soda, which is usually treated and carbonated municipal water, went after the premium bottled water market with a vengeance. Some club sodas preferred to be known as seltzers. Several hotels and restaurants began to offer "bottled mineral water" under their own private label. Private labelling is often little more than carbonated municipal water. Confusion surfaced in the consumer's mind and the FDA seemed equally perplexed over bottled water nomenclature.

But the best was yet to be. In 1979, Roland Au of Santé in Northern California introduced flavored mineral water. Since people were adding a slice of lime to their Perrier, why not add the essence of lime, and later other flavors, in the actual bottling process? Here was the rebirth of the soft drink without sugars, juices or artificial ingredients. The flavor essence was little more than just a hint, but enough to establish an explosive offspring of premium bottled water. Other bottlers saw the sometimes six-to-one selling ratio of flavored water over nonflavored and dived into the market. Crystal Geyser introduced Colaberry and three citrus flavors in 1982. Calistoga followed in 1983, and Perrier took the plunge shortly thereafter. Flavored water became a natural extension of many bottled water companies, and the only thing in short supply was shelf space.

Arrowhead spearheaded another variation of flavored water by adding actual fruit juice to carbonated water. Using real lemon, lime and orange fruit juice, their flavored waters packed a punch and stimulated both the olfactory and taste sensors.

Flavored bottled waters' 20 percent growth rate in 1986 continued to impress the business world with their potential. More than $250 million in sales of flavored seltzers was forecast for 1987 by *Beverage Digest*, a leading industrial trade magazine. Americans in ever-growing numbers are claiming them as their preferred recreational drink, the "fastest-growing soft drink since the introduction of caffeine-free colas, according to industry figures. "The mystique of bottled water is definitely drawing people to flavored sparkling water," Jeanne Lukasik of *Beverage World* commented at the 1986 Conference of the National Soft Drink Association, concluding, "Flavored bottled water is assuming the role of the cocktail."

Stroh Brewery's Sundance Natural Juice Sparkler along with Australia's Koala Springs used sugars and fruit juice, making a big splash as a soft drink alternative. Yet, these beverages are anything but bottled water. In the purist form, bottled water should have nothing added to it nor taken away. The words *spring* and *mineral water* create confusion when used with soft drink beverages, often leading the consumer to believe that the beverage is noncaloric and natural.

The reigning king of bottled water is Fred Sipper, of M&M Distributing in New York. With 87 brands of bottled water from Evian to San Pellegrino, Mr. Sipper is responsible for up to 70 percent of some water company sales in the East Coast.

But is all that sparkles golden? Would billions of bubbles bring millions of dollars? Many dived headfirst into the water business and never surfaced again!

Ken Sheffield's O_2 burst onto the California water market as a sobering sister product for its renowned winery, Chateau St. Jean. Originally positioned as a premium carbonated water, it also hinted that it could cure hangovers. A perfect marriage for a wine company, but somehow the wine distribution channels didn't work. The water product sank and now the very mention of Sheffield's O_2 around Chateau St. Jean brings out some sour grapes.

Manitou's Rick Sajbel has struggled to keep afloat in the premium carbonated water business and has borrowed money from the government, employees' parents and even his 87-year-old piano teacher! Even so, there are weeks when he can't sell his mineral water, not for lack of demand, just lack of glass — because he can't pay for it.

Ty Hill could turn water into wine with his strong beliefs and personal convictions. Mendocino Mineral Water was his mecca and his cozy bottling plant surrounded by redwoods captures the romance of bottled water. Ty and his wife, Gayle, started bottling water in 1982. They mortgaged their ranch, borrowed money from the bank and plowed every last cent into Mendocino to keep it alive. They never saw a profit until they sold it to Vinica of Lucerne, Switzerland, in 1986.

Bottled water buy-outs are the one place where the really big bucks are being made. The analysts who predicted fiscal fireworks when Beatrice sold off its bottled water division were not disappointed. Arrowhead, with the rest of the Beatrice package, sold to The Perrier Group for over $400 million, the biggest bottled water sale in history.

And what is the next step in bottled waters? Crystal Geyser may have found it with its Vitafort, an expermimental beverage that is a licensed nutritional supplement developed for the bottling firm by Hoffman La Roche. Seven water-soluble vitamins, including B complex and C, were formulated for Crystal Geyser's carbonated mineral water. One ten-ounce serving provides 50 percent of the recommended daily allowance of those vitamins. Vitafort does not contain added color, odor, taste or calories, containing only sparkling water, minerals and vitamins. It could be advertised as a glassful of what's good for you. The big question is, will the public go for it?

Bottled water in America is nothing new. Most bottlers were established well over fifty years ago — some even over a hundred. Nevertheless, many Americans have still not learned to accept the fact that city water is often undrinkable, and scoff at the idea of paying hard-earned money for water. "I suppose they sell air, too!" is the standard response.

Certain regions are known as bigger markets for bottled water than others. Much depends on the quality of the local tap water. In mountain areas, for instance, people do not buy much bottled water because their mountain municipal water is usually very good. The decision to drink bottled water may have nothing to do with the actual quality of tap water, and instead depend entirely on flavor—which in turn depends on temperature. Most tap water is acceptable if it is cold enough. In most of the colder regions of the United States municipal water pipes are laid deep in the ground to avoid freezing in winter. At such depths the water stays cool all year. As water cools, unpleasant-tasting minerals and elements become less detectable, making cold tap water taste decent—whatever it contains.

In warmer areas, such as Southern California and Florida, there is little danger of frost, so water pipes are laid close to the surface since digging deep trenches is costly. Consequently, water comes out of the taps warm, and tastes bad.

The result is that on the East Coast, only one family in 2,000 drinks bottled water regularly, while California accounts for 50 percent of all national bottled water sales.

Many bottled water companies are still family businesses, although some have been bought up by large conglomerates. A few have ventured into related concerns such as the manufacture of water purification equipment and water coolers.

Domestic bottling companies usually supply water in five gallon glass or plastic bottles, which must be used with a dispensing cooler unit. A typical household uses three five gallon bottles per month, paying about 50 cents a gallon. Several kinds of dispensers are usually available, both electric and mechanical. An electric cooler rents for about four dollars a month, and a non-electric one, which cools through evaporation to keep the water about fifteen degrees below room temperature, for about 60 cents. Some companies sell hand pumps and tilt stands which cost less than ten dollars. In some cases water bottlers are happy to break even on the sale of their water—their profit comes from dispenser rentals.

Waters for human consumption may come from a variety of sources and undergo a variety of processes before bottling. Generally, four basic kinds of bottled water are sold. First, there is *plain drinking water,* which may come straight from a spring or a well, or it may be *scientifically purified with minerals added to improve the taste.* This drinking water is sometimes available with added fluoride, and offered to families with children to help retard tooth decay. Also available are *purified water* and *distilled water. Purified water,* sometimes referred to as man made water, is almost unique to the United States. It is usually nothing more than municipal water which has had all the minerals and elements, both good and bad, removed. This is accomplished by distillation, deionization, reverse osmosis, and/or electrodialyis. Purified water has usually been chlorinated at some point, and the chlorine also removed by filtration.

Distilled water is the closest thing we have to pure water. This is the state of water in clouds, before it falls to earth as rain. Distilled water is produced by heating water into vapor and allowing it to condense into liquid form again, leaving it free of most solids, minerals and trace elements. Some doctors prescribe it for patients on low sodium diets. At Murietta Hot Springs, Dr. Reems has patients drink distilled water and lemon juice in a program designed to detoxify the system. He follows this by having patients drink small amounts of natural spring water, complete with minerals.

Scientific research has shown that minerals in water, as in foods, do contribute to health. In Chapter 3 we looked at some findings on disease incidence in hard and soft water areas, and saw a positive correlation between heart, liver and bladder disorders and drinking water low in minerals.

Thomas Love of the Washington, D.C., *Star* reported on November 13, 1974, that Dr. Raymond A. Saunders, who performed a study of Washington's municipal water systems, believed distilled water made from municipal water systems would contain a higher concentration of carcinogenic substances than the tap water itself. "Most of the substances we found in the Washington water have a lower boiling point than water," he explained. Thus carcinogenic substances could enter the vapor stage earlier than water and wind up being more concentrated in the end product. Dr. Robert Harris of the Environmental Defense Force, studying a link between carcinogens in drinking water and cancer deaths in Louisiana, agreed with Saunders. The interaction between chlorine and organic compounds is a major cause of carcinogens.

Paul Bragg preferred to ignore these findings. An enthusiastic advocate of distilled water, he wrote several books about drinking water. In them he made some extravagant claims for the benefits of distilled water. He claimed it could prevent hardening of the arteries, since, according to him, all other waters cause mineral buildups in tissue which cannot be dissolved by the body. Bragg's book, *The Shocking Truth About Water,* inspired many readers to purchase a particular brand of water distilling system he recommended. His enthusiasm was perhaps explained by the fact that he owned the company which manufactured these units.

Transportation Problems

Even if water is bottled in a plant directly over the spring, a bottling company still has to carry it to the customers, and transportation of bottled water presents several problems.

Bottled water must be free from microorganisms. To achieve this most water bottlers have chosen the process of ozonization, since ozone is not soluble in water and can be used in maximum quantities, in contrast to chlorine. Ozone is oxygen in an unstable form, having three atoms to one molecule instead of two. It will oxidize organic matter in water, completely removing undesirable odors, tastes, and so on. By injecting O_3 bubbles through the water, sterilization is achieved in just a few minutes. Ozonization also sterilizes the bottle and cap. Ozone has virtually no color, odor or taste and escapes from the bottle upon opening.

Although bottled water in five gallon bottles leaves the plant fairly pure and free of bacteria, the bottles are transported in open trucks. By the time they are delivered to the customer there may be some dirt and grime around the neck of the bottle. Most companies recommend the bottle neck be wiped with a cloth which has been dampened with disinfectant before placing the bottle in the cooler.

Some naturally sparkling, or carbonated, waters when bottled and transported, will arrive at their destination with loss of carbonation. For example, sparkling Vichy Célestins has far more effervescence in France than it does by the time it arrives in Los Angeles. Certain other waters, like Perrier, seem to retain their sparkling characteristics even after long periods of storage.

Water high in sulphur content, after being bottled for some time, may begin to smell more and more of rotten eggs. It has been discovered that such water is actually poisonous and prolonged use of it may actually kill.

Mineral water that is well bottled can remain drinkable for decades. John Scott, of Health Waters, recently opened a bottle of Mountain Valley which had been left in storage for over 50 years. He reported that it was delicious, and, if anything, improved with age.

In Quest of the Perfect Spring

"If there's magic in the world
it must be in water."
— Loren Eisley

There is something magical and mysterious about a spring. Water from deep within the earth surfaces, bubbling over rocks, pushing its way through dense growth, or curiously springing up in the midst of a barren desert. Wherever it surfaces, water brings life to micro and macro organisms, which exist only with water. And each spring has its own identity, as unique and individual as a fingerprint.

Springs come steaming hot and icy cold; they bubble with carbonation, smell of sulphur, color the soil red, white or blue; brave a thousand years and then disappear without a trace. There are springs that burst from the earth as torrent. Rivers and springs that barely seep. Springs may taste sweet or sour or salty. Some sparkle with clarity; others are darkened with turbidity. Yet all share one common feature: they come to us from deep inside the earth without any help from man or machine.

Man has always sought out springs of perfection, waters that made him feel more alive, healthier, vibrant: waters that rejuvenate. Ponce de Leon was searching for a spring. Early man would travel far to camp by such springs; Greeks and Romans built temples around them; Indians made them part of their rituals. At the turn of the century, one geologist, Gerald Waring, combed the world to log and describe all the springs he could discover. Many have followed in his footsteps, and a few have found at the end of their quest a perfect spring whose waters belong in a bottle.

What makes a spring good enough to bottle? First, it must surface in an environment free from contamination. Heavy industry, active agriculture, or populated cities are not good settings for pure water. But in the country, where the air is fresh, and where rolling hills or mountain peaks protect springs from the onslaught of civilization, great water can emerge. More than pure, the water must have a balanced analysis. Too much sodium, too little magnesium or calcium, makes the difference between the ordinary and the extraordinary spring. And its water must remain consistent. If the analysis changes from day to day, or even from season to season, it suggests that the spring is recharged by surface groundwater which may be contaminated from the air or land. If the flow is not consistent, this, too, indicates possible intrusion.

A perfect spring does not need any kind of treatment or purification. Natural spring water has nothing added to it and nothing removed. Its waters should be bottled at their source.

All still water contains bacteria in its natural state. But there is bacteria and bacteria. Yogurt and great cheeses depend on good bacteria, while doctors depend on bad bacteria! Many feel that the good bacteria in natural spring water is as beneficial as the bacteria in yogurt. In much of the world, bottled water is not sterilized with ozone, chlorine or other processes that kill bacteria, because the bacteria or flora in the water source is good. In the United States, bottlers are compelled by law to sterilize their water, however good its bacteria.

Carbonation, whether naturally occurring in the spring or during the bottling process, kills bacteria. Springs whose waters bubble like champagne are free from bacteria.

A perfect spring for bottling water, although removed from civilization, must be accessible to its markets. No matter how ideal the spring we might discover at Basecamp Four on Mt. Everest, transporting its waters from the Tibetan Plateau's 26,200-foot heights to the 200 block of Beverly Hill's Rodeo Drive would sink a bottling business.

I have searched across the United States for springs and have found very few new ones good enough to bottle. Most of the great ones were already being bottled; many are featured in Part Two of this book, the Encyclopedia of Bottled Waters. The few which still are out there waiting to be discovered may one day turn up on your table and in a revised edition of H_2O.

AN ENCYCLOPEDIA OF BOTTLED WATERS FROM AROUND THE WORLD

Information in the water analysis tables that follow was provided by the bottlers, although a few companies elected not to provide a water analysis. We have rearranged elements in the tables for uniformity and to offer a convenient comparison between waters.

Calcium, magnesium, sodium and potassium are listed first, following the almost universal practice of the bottling companies themselves. Single elements follow in alphabetical order. Radical groups, such as bicarbonates, sulphates, or silicates, are listed after the elements; miscellaneous information, such as total dissolved solids and pH, has been listed last. The symbol < has been used in preference to the words "less than."

ARIZONA

Arizona Sparkling Bottled Water

(Phoenix)

Arizona Sparkling Bottled Water was established in 1960 by Thomas Gallagher, soon joined by Howard Huber and Bud Willard, who then formed a partnership. The water is drawn from a well in northwest Phoenix, located on company property. Judging from the direction of the flow, the source itself is fed from the mountains north of Phoenix. The well is 800 feet deep and sealed in a concrete sleeve to eliminate any possible contamination. The water is so pure it is not chemically treated, processed or fluoridated.

Arizona Sparkling Water Analysis

	Parts per Million
Calcium (Ca)	4
Magnesium (Mg)	1
Sodium (Na)	71
Chloride (Cl)	43.0
Chromium (Cr)	.02
Fluoride (F)	.6
Iron (Fe)	.05
Sulphates (SO_4)	6
Nitrates (NO_3)	2.6
Total Dissolved Solids	200
pH (Hydrogen Ion Concentration)	8.9

Crystal Bottled Waters Company

(Phoenix)

The company was founded in 1948 when three Arizona water operations — Mountain Spring Water Company, Westwood Water Company and Artesia Water Company — banded together. The three principals were Jim Williams, Glenn Jones and Charles Christopher. Ten years later, Crystal Associates (Williams and Christopher) purchased the firm and incorporated. McKesson acquired the firm in 1969, with Crystal becoming an important unit of its water division. A $1.1 million production plant was opened in 1974.

Crystal Bottled Waters

	Parts per Million
Calcium (Ca)	.7
Magnesium (Mg)	.8
Sodium (Na)	4.8
Chloride (Cl)	3.4
Fluoride (F)	.8
Bicarbonates (HCO_3)	5.8
Sulphates (SO_4)	3.4
Total Dissolved Solids	19.8
pH (Hydrogen Ion Concentration)	6.8

Crystal Bottled Waters starts with natural well water, passes it through a reverse osmosis unit, then de-ionizes it. Crystal-purified is a mineral-free water for all distilled water uses. To prepare Crystal-Fresh drinking water, selected minerals, including fluoride, are added to the purified water.

The new Phoenix plant is located over one of the area's best underground water sources. Nearly, 1,000 feet deep, the well pumps 100 gallons of water a minute. Crystal-Purified and Crystal-Fresh are available in five gallon containers for delivery to homes and offices, and in gallon containers in retail outlets.

ARKANSAS

Caddo Valley Spring Water

(Hot Springs)

Caddo Valley Spring waters, high in calcium and low in sodium, may have been sampled by Hernando de Soto while he was battling with the valley's Tula Indians in 1541. The springs, along with 40 acres of surrounding land, were homesteaded by William J. Cox just after the Civil War. The present owner, Mrs. E. Harnik of Hot Springs, keeps alive the tradition that those coming to the springs may take away free as much water as they can carry. And people do come with plastic five-gallon containers and Gallo wine jugs. In the old days it was fruit jars and galvanized tubs.

That openhanded attitude has generated a flourishing business; sales have increased every year for 26 years. In addition to supplying its local customers, millions of gallons are shipped to Texas, Oklahoma and Kansas, where it is sold under several different labels.

Mountain Valley Water

(Hot Springs)

Mountain Valley Water has a long and colorful history. According to the records of Hernando De Soto's 1541 exploration, the Spaniard found Indians of many tribes peacefully sharing recuperative waters at the hot springs in what is now Arkansas. Indians and explorers were followed by French trappers and traders early in U.S. history. President Thomas Jefferson set up a commission to study the waters, and in 1832, as a result of their report, Congress reserved land containing 47 springs for the federal government. After the Civil War, "Lockett's Spring" was acclaimed locally. In 1871, John Green, a pharmacist, purchased Lockett Springs, miles from the center of Hot Springs, and began to sell the water as Mountain Valley Mineral Water. By 1879, Mountain Valley Water was being enjoyed by residents of many states, according to letters published by the *Hot Springs Illustrated News.*

For 3,000 years spring waters achieved popularity through their acclaimed therapeutic benefits. Mountain Valley joined the parade of "cure waters," and for the next 80 years was advertised as a therapeutic aid. In 1879 John Green and his brother Peter sold the springs to W. N. Benton of St. Louis. By 1882 Mountain Valley was sufficiently well known that a group chartered the "Hot Springs and Mountain Valley Railroad" to bridge the 12 mile gap between Hot Springs, bustling with hotels and bathing houses, and the increasingly famous cold water spring, which by then had a hotel and cottages.

Mountain Valley received a corporate charter from the State of Arkansas in 1882 "to ship . . . and sell . . . in all parts of the world." In 1894 demand was great enough to open a franchise in Philadelphia. A second franchise opened in New Orleans at about the same time. These were among the first franchises issued by any business, certainly the first ever issued to an Arkansas firm.

A few years later a surgeon with enormous influence in the medical world, Dr. J. B. Murphy, of Chicago, swore to his friends that Mountain Valley Water was the finest in the world. One friend, August Schlafly of Carlyle, Illinois, was convinced enough to pur-

Mountain Valley
Recent Water Analysis

	Parts per Million
Calcium (Ca)	69.68
Magnesium (Mg)	10.56
Sodium (Na)	2.88
Potassium (K)	1.13
Aluminum (Al)	.70
Bromide (Br)	trace
Chloride (Cl)	6.38
Iodide (I)	trace
Iron (Fe)	.027
Lithium (Li)	trace
Manganese (Mn)	trace
Bicarbonates (HCO_3)	238.05
Carbonates (CO_3)	6.99
Sulphates (SO_4)	9.72
Nitrites (NO_2)	.47
Nitrates (NO_3)	0
Phosphates (PO_4)	.25
Ammonium (NH_4)	$<.05$
Silica (SiO_2)	12.08

Mountain Valley
Water Analysis
Three Decades Ago

	Parts per Million
Calcium (Ca)	68.04
Magnesium (Mg)	7.50
Sodium (Na)	3.40
Potassium (K)	1.42
Aluminum (Al)	.13
Bromide (Br)	trace
Chloride (Cl)	5.00
Iodide (I)	trace
Iron (Fe)	1.29
Lithium (Li)	trace
Manganese (Mn)	.02
Bicarbonates (HCO_3)	240.34
Carbonates (CO_3)	—
Sulphates (SO_4)	8.89
Nitrites (NO_2)	.10
Nitrates (NO_3)	trace
Phosphates (PO_4)	.25
Ammonium (NH_4)	.04
Silica (SiO_2)	13.75

chase the spring in 1902 with two other doctors, whose interest he bought out in 1906.

Dr. Murphy remained so keenly interested in Mountain Valley Water that he considered buying the spring himself, but was dissuaded when he became President of the American Medical Association. The Association felt its members should not involve themselves in commercial ventures.

Nineteen-eight was the greatest single year for Mountain Valley's expansion—franchises opened in Chicago, New York, Indianapolis, and other cities. Since Dr. Murphy was practicing in Chicago, the franchise there was an immediate success. Larry McNally, son of the original manager, continues as head of the franchise today.

The New York franchise began very differently. Two men, both affluent New Yorkers who did not know each other personally, were returning home from Hot Springs. In the dining car of the train each produced a bottle of Mountain Valley for his table. A discussion about the water followed, and they agreed a company should be formed to make it available in New York. "Let's each put up fifty percent of the cost," one suggested. The other agreed. They shook hands, exchanged cards and learned each other's names for the first time. One was William Randolph Hearst, the crusading publisher. The other was Richard Canfield, New York's most infamous gambler, against whom Hearst newspapers were conducting a campaign! The two men went ahead with their joint venture nevertheless, and the first carload of bottled Mountain Valley was shipped from Hot Springs to New York, each bottle on its side with a long cork in it, in the manner of wine.

When the shipment arrived, Canfield and Draper (manager of the new office) were surprised to find thousands of empty bottles. The water had disintegrated the corks, an early sign that Mountain Valley water was an excellent solvent. In 1921 an eastern bottling plant was opened. Mountain Valley supplied water in drums until 1939, when the company began using glass lined tank cars. Today, although many water companies use plastic containers, Mountain Valley still uses glass bottles, since the water picks up taste and odor from every plastic tested so far.

In the 1920s the first Mountain Valley shipments were made to the United States Senate. The President also became a customer. (In 1956 President Eisenhower told a press conference, following a heart attack, that he drank Mountain Valley Water on the advice of his physician.)

A fully protected spring should have its aquifer, from which all its water is ultimately drawn, protected from the pollution of factories, manufacturing, cities, and even farming. Timberland such as this which protects Mountain Valley is an ideal aquifer.

Through the years therapeutic claims made for mineral waters have moderated, with the blessing — and at times urging — of public authority. As early as 1914 the Federal Trade Commission rapped the knuckles of two exuberant Philadelphia distributors who had advertised their water as a cure for every known disease. In 1956 the Food and Drug Administration attacked Mountain Valley in court. However the company produced convincing evidence of the water's therapeutic merit, gathered from hospital and clinical work and laboratories in Houston, St. Louis, Cincinnati, Chicago, Philadelphia and New York. The court found in favor of Mountain Valley. The Kefauver Amendments to FDA regulations in 1964 contained a "grandfather clause" which permitted Mountain Valley to continue advertising its product as of therapeutic value.

The complexion of the market was changing, however. Old fashioned advertising of spring water as an aid for arthritis and kidney conditions was giving way for several reasons. Science and medicine were able to use specific remedies for these diseases instead of depending on the general empirical benefit of a water like Mountain Valley. In addition, more doctors were recommending spring water for its low sodium content, natural hardness, purity and good taste. On top of this, America had developed a desire for good tasting water. Mountain Valley headed the list in fulfilling this new need.

Not all Mountain Valley's customers are human. Famous race horses have drunk it, including 19 of the last 23 winners of "horse-of-the-year" honors. Among the most famous thoroughbreds who have trained with Mountain Valley were Gallant Fox, Bubbling Over, Nashua, Bold Runner, Sword Dancer, Gallant Man, Secretariat and Kelso.

Beneath the spring lies a huge marble bed millions of years old. Surrounding it are 500 acres owned by Mountain Valley directly adjoining a timberland preserve. A glass dome covers the spring, reducing chances of air pollution. Mountain Valley water contains no added chlorine or other chemicals; it is moderately hard and mildly alkaline with low sodium content, and flows at a constant temperature. (Water temperature which does not vary is a useful rule-of-thumb indication of spring purity, since it is assumed that water which is unaffected by ambient temperature is deep enough to prevent chemical contamination or dilution with surface water.)

In April of 1987, Mountain Valley was purchased by Sammons Enterprises, Inc., of Dallas, Texas. Sammons also purchased Carolina Mountain Water Company and Diamond Water of Hot Springs.

Ozarka Water Company

(Hot Springs)

The Tula Indians, who fought against De Soto and his men at Caddo Valley in 1541, were using the water of the crystal spring on the west side of the Ozark Mountains of Arkansas.

In the 1920s the Ozarka Water Company was formed to collect the water from this protected, historic spring. Ozarka has become the leading bottled still water brand in Texas and the drinking water of the famed Dallas Cowboys, the Houston Oilers, and the Houston Astros professional sports teams.

Ozarka mountain spring water, drinking, and distilled water are delivered to homes and offices throughout Texas.

Ozarka also offers a popular sparkling mountain spring water flavored with a squeeze of juice, widely available in supermarkets. Recently the company was acquired by The Perrier Group.

CALIFORNIA

Alhambra Crystal-Fresh Water Analysis

	Parts per Million
Calcium (Ca)	.7
Magnesium (Mg)	.8
Sodium (Na)	3.8
Chloride (Cl)	3.5
Fluoride (F)	0
Bicarbonates (HCO_3)	5.8
Carbonates (CO_3)	0
Sulphates (SO_4)	3.4
Nitrates (NO_3)	0
Total Hardness, as $CaCO_3$	5.1
pH (Hydrogen Ion Concentration)	6.8

Alhambra National Water Company

(San Francisco)

The company dates back before the great San Francisco earthquake of 1906, when, as the National Water Company, it began producing and selling distilled water. Acquiring the Alhambra Water Company of San Francisco in 1945, it changed its name three years later to the Alhambra National Water Company, Inc. Currently, it is a subsidiary of the McKesson Company.

Three products — Purified, Crystal-Fresh and Fluoridated waters — are sold by the company. Purified water is obtained by passing municipal water through a reverse osmosis and deionization process, sterilizing and demineralizing it. Minerals are added to Purified by formula to obtain Crystal-Fresh. Fluoride *and* minerals are added to Purified to obtain Fluoridated water. These products are marketed in the Bay Area and through branch distribution centers in Hayward, Salinas, Santa Clara, Santa Rosa, and Vallejo, by delivery to homes and offices in five-gallon bottles.

Arrowhead Puritas Water Company

(Los Angeles and Orange County)

Long before Spanish explorers arrived in North America, Indians drank and bathed in the therapeutic hot springs at Lake Arrowhead. In the early 1880s David Smith came to the San Bernardino area from Ohio and recognized the value of the springs. He built a health spa at the hot springs and called it the Arrowhead Sanitarium after the large natural arrowhead formation still to be seen on a mountain above the area.

Visitors from all over the United States came to take the waters, and many wanted to have the water shipped back to their home towns. So the water was bottled at the spring and shipped out by electric railway, but one plant could not meet the tremendous demand. In 1909 a second plant was established in Los Angeles to bottle the water which arrived from the spring by rail in tank cars. Deliveries of one gallon to five gallon bottles were made by horse drawn wagons in the Los Angeles area.

Water distilled in Los Angeles was used by an ice company to make crystal clear ice. In 1894, after many requests to use the distilled water for drinking, it was bottled and sold under the Puritas name. Arrowhead and Puritas merged in 1929 to become one of the largest water bottlers in Southern California. The company bottles spring, distilled and fluoridated water, and manufactures electric coolers, paper cups and coffee, soup and beverage mixes. The division which manufactures water conditioning equipment also provides water treatment units for industry.

In 1987, The Perrier Group purchased Arrowhead from Beatrice.

Arrowhead Spring Water Analysis

	Parts per Million
Calcium (Ca)	15.2
Magnesium (Mg)	2.7
Sodium (Na)	9.4
Potassium (K)	1.5
Arsenic (As)	none detected
Barium (Ba)	.005
Cadmium (Cd)	none detected
Chloride (Cl)	7.5
Copper (Cu)	.012
Fluoride (F)	.12
Iron (Fe)	.015
Lead (Pb)	none detected
Manganese (Mn)	.0005
Mercury (Hg)	.0002
Selenium (Se)	none detected
Silver (Ag)	none detected
Strontium (Sr)	.35
Bicarbonates (HCO_3)	67.1
Sulphates (SO_4)	2.5
Nitrates (NO_3)	.8
Phenols	none detected
Total Dissolved Solids	103.5

Calistoga Mineral Water Company

(Calistoga)

The Wappo Indians first discovered Calistoga's "Old Faithful" geyser, soaking in its hot mineral pools to cure their ailments. Later, Calistoga was developed as a spa town by Sam Brannan, an ardent health spa enthusiast who was said to be California's first millionaire, thanks to the gold rush.

Calistoga Mineral Water Analysis

	Parts per Million
Calcium (Ca)	80
Magnesium (Mg)	3.4
Sodium (Na)	128
Potassium (K)	0
Chloride (Cl)	80
Fluoride (F)	1.3
Iron (Fe)	1.0
Manganese (Mn)	.06
Bicarbonates (HCO_3)	220
Carbonates (CO_3)	0
Sulphates (SO_4)	15
Nitrates (NO_3)	.80

Alkalinity	180
Total Hardness, as $CaCO_3$	20
Total Hardness, as $MgCO_3$	14
Total Dissolved Solids	750
Turbidity	.5
pH (Hydrogen Ion Concentration)	5.3

In 1924, the Calistoga Mineral Water Company was formed to bottle Calistoga waters. In 1983, Calistoga was one of the first companies to begin marketing sparkling mineral waters flavored with the natural essence of fruits. Four years later, Calistoga introduced "Sparkling Juice," a refreshing combination of its mineral water and pure fruit juice in a wide range of flavors from cranberry to passion peach.

The Perrier Group purchased Calistoga, and in 1983, introduced flavored waters. In 1987, Calistoga began marketing natural juices blended with mineral water.

Crystal Geyser Water Company

(Calistoga)

Crystal Geyser Water Analysis

	Parts per Million
Calcium (Ca)	12
Magnesium (Mg)	3.1
Sodium (Na)	130
Potassium (K)	8.7
Fluoride (F)	< 1
Bicarbonates (HCO_3)	170
Nitrate (NO_3)	< 0.1
Total Dissolved Solids	590
pH (Hydrogen Ion Concentration)	5.2

Bottled mineral water from the slopes of Mt. St. Helens, Crystal Geyser Water comes boiling hot from subterranean strata. It is pumped and cooled to 35 degrees (by a specially designed water-handling system that protects the water's natural purity) carbonated, and bottled at the source.

The young company was started by two young ex-bankers, Peter Gordon and Leo Soong, who scoured the Napa and Sonoma country in consultation with geologists looking for the "perfect water," before setting up Crystal Geyser Water Company. The partners claim the purity of their water allows Crystal Geyser to hold its carbonation when opened or poured over ice longer than other mineral waters.

"Just as the wines of Napa Valley have come to be recognized as the equal of the best French wines, so will the waters of Napa Valley gain recognition in relation to their French counterpart," Soong predicts. "And," adds Gordon, "just as each wine has its own taste, each water tastes different because of the differences in mineralization. We think ours is the best tasting of all."

Deep Rock Water Company

(Los Angeles)

Mr. and Mrs. Manuel Rattner founded the Deep Rock Water Company in 1929, drawing the water from two 300 foot artesian wells on their property. Paul and Evelyn Pendleton took over Deep Rock in 1947, and today the company is still run by their sons, Larry and Don Pendleton. A new well, now the source of most of the water, was dug in 1974.

The company also bottles distilled and purified water. The distilled water is produced by vapor compression, and the purified water by deionization and reverse osmosis.

Supermarkets in Southern California sell Deep Rock water in one gallon glass bottles. It is available in five gallon bottles for home delivery.

Deep Rock is part of the Hinckley & Schmitt group.

Deep Rock Water Analysis

	Parts per Million
Calcium (Ca)	81
Magnesium (Mg)	16
Sodium (Na)	55
Potassium (K)	2.0
Barium (Ba)	<.01
Boron (B)	20
Cadmium (Cd)	<.01
Chloride (Cl)	65
Chromium (Cr)	<.01
Copper (Cu)	.07
Fluoride (F)	.68
Iron (Fe)	.02
Lead (Pb)	<.01
Manganese (Mn)	<.01
Mercury (Hg)	<.001
Silver (Ag)	.01
Zinc (Zn)	.21
Bicarbonates (HCO_3)	214
Carbonates (CO_3)	0
Sulphates (SO_4)	87
Nitrites (NO_2)	.05
Nitrates (NO_3)	19
Silica (SiO_2)	515
Phenols	<.001
Total Dissolved Solids	265
pH (Hydrogen Ion Concentration)	7.7

Forest Lake Spring Water Company

(Lake County)

In 1940, a retreat on Cobb Mountain about 90 miles from San Francisco claimed to have a wonderful water source. From all over California people came to drink the water and relax in a beautiful, unspoiled environment. Calso Spring Water Company began bottling the water and selling it under the name Calso water. Calso Water Company sold the business to Forest Lake in 1974; Calso moved to Menlo Park where it still sells bottled water.

Forest Lake spring water comes to the surface at 42° C, and has a total dissolved solids count of only 27 ppm in its natural state. It flows at a rate of 150,000 gallons per day.

The balance of this water is excellent. The Finnigan Corporation, a major producer of water analysis equipment for the Water Quality Laboratory, wrote, "Our analysis shows that your water sample is one of the purest we have seen in our years of analyzing water samples."

Forest Lake Spring Water Water Analysis

	Parts per Million
Arsenic (As)	<.01
Barium (Ba)	<.1
Cadmium (Cd)	<.005
Chloride (Cl)	1.6
Chromium (Cr)	<.005
Copper (Cu)	<.05
Fluoride (F)	<.32
Iron (Fe)	.02
Lead (Pb)	.01
Manganese (Mn)	<.05
Mercury (Hg)	.002
Selenium (Se)	<.005
Silver (Ag)	<.01
Zinc (Zn)	<.005
Sulphates (SO_4)	.5
Nitrates (NO_3)	.1
Phenols	.05
Total Dissolved Solids	27

A. The Sierra Nevada Range. B. Snow melts, passing through layers of sedimentary, volcanic and basement rock. C. The Great Valley, a massive expanse of sedimentary rock that stretches over 100 miles. D. Lodestone, a giant rock containing highly magnetic qualities. E. Cobb Mountain, an extinct volcano made up largely of volcanic and sedimentary rock. F. The Spring.

As melting snow travels down through the mountains, filtering through layers of volcanic and other sedimentary rock, it begins to collect in underground tributaries. The water continues its journey, converging to form an underground river. At the base of the mountain the water enters the Great Valley and passes through innumerable beds of sand and silica, each layer filtering the water further. Still underground, the river passes through a deposit of lodestone and more impurities are drawn off by the natural magnetic qualities of this giant rock. After 212 miles the river reaches Cobb Mountain. Here, water pressure forces the water thousands of feet upward through many layers of volcanic rock. As the water passes through each layer, it is filtered over and over, until it reaches the mouth of Cobb's Spring and bubbles forth.

Mendocino
Mineral Water
Water Analysis

	Parts per Million
Calcium (Ca)	380
Magnesium (Mg)	120
NaCl (Mineral salt)	105
Potassium (K)	3
Fluoride (F)	.34
Iron (Fe)	2.8
Lead (Pb)	.01
Manganese (Mn)	.23
Zinc (Zn)	.08
Bicarbonates (HCO_3)	1890
Nitrates (NO_3)	.5
Sulphates (SO_4)	1
Total Dissolved Solids	2080
pH (Hydrogen Ion Concentration)	6.3

Mendocino Mineral Water Corporation

(Mendocino County)

Mendocino Mineral Water is one of the few bottled waters in the United States that contains its own natural carbonation and its own natural pure oxygen. Its mineral content, a total dissolved solids of 2,080 ppm, is one of the highest and most balanced of any mineral waters sold. The artesian source, from which it flows at the rate of 60 gallons per minute, is protected by hundreds of acres of forest. Totally bacteria-free as it gushes bubbling from the earth at a constant yearly temperature of 53°F, the waters were originally discovered and used by the Indians of the Bay region, and later by early settlers and the Gold Rush's forty-niners.

This naturally carbonated, highly mineralized water surfaces on a 197-acre ranch. It was once bottled in 1880 by G.M. Henderson, but operations ceased before the turn of the century. The property was purchased by Ty Hill in 1970, who established Mendocino Mineral Water Corporation in 1980. He began to market the water in 1982. The company was purchased by Vinica of Lucerne, Switzerland, in 1986.

Mountain Spring Water Company

(Santa Barbara)

Santa Barbara, famous for its sulphur hot springs, also has some well balanced drinking water springs in the Santa Ynez mountains. Since the turn of the century Mountain Spring has been bottling water from one spring in the Painted Cave region. The spring, protected by 40 acres of land, surfaces through a sandstone bed and is bottled at the source. The water, in five gallon bottles, is transported down the mountain and distributed through a three-county area.

Mountain Spring Water Analysis

	Parts per Million
Calcium (Ca)	39
Magnesium (Mg)	7.2
Sodium (Na)	16
Potassium (K)	.5
Chloride (Cl)	30
Fluoride (F)	.2
Iron (Fe)	.01
Manganese (Mn)	.01
Bicarbonates (HCO_3)	108
Sulphates (SO_4)	26
Nitrates NO_3)	2.0
Silica (SiO_2)	12
Free CO_2 gas	13
Total Hardness, as $CaCO_3$	125
Total Dissolved Solids	187

A Magic Fountain

Indeed is that which through its waters cures Bright's and other Diseases of the Kidneys, Dyspepsia, Rheumatism, Dropsy, etc., and this is what the celebrated

KALIUM SPRING WATER

(Registered)

Will do. Do not take our word for it, but send for pamphlet containing the voluntary testimonials of those who have tried it.

Remember, all you have to do is to drink FREELY of a pure, delightful water, without taste or smell.

IT IS CHEAP and ABSOLUTELY EFFECTIVE.

Orders by mail promptly attended to.

3 Miles on the Camden and Atlantic Railroad.

Address WILLIAM JONES, (at the Spring), Collingswood, N. J.

N. B.—Recollect, we have no agent; wagon in town every day.

Advertisement of Kalium Spring Water

(From the "Camden Daily Courier" of July 29, 1886.)

Napa Valley
Springs
Water Analysis

	Parts per Million
Calcium (Ca)	2
Magnesium (Mg)	1.9
Sodium (Na)	178
Potassium (K)	11
Arsenic (As)	<.001
Barium (Ba)	<.04
Cadmium (Cd)	<.009
Chloride (Cl)	209
Chromium (Cr)	<.006
Copper (Cu)	.019
Fluoride (F)	.06
Iron (Fe)	.02
Lead (Pb)	<.001
Manganese (Mn)	.01
Mercury (Hg)	<.001
Selenium (Se)	<.001
Silver (Ag)	<.009
Zinc (Zn)	.026
Bicarbonates (HCO_3)	156
Carbonates (CO_3)	0
Sulphates (SO_4)	9.7
Nitrates (NO_3)	.04
Cyanide (CN)	<.01
Silica (SiO_2)	110
Alkalinity, as $CaCO_3$	128
Total Hardness, as $CaCO_3$	13
Total Dissolved Solids	619
Color	1
Turbidity	.4
pH (Hydrogen Ion Concentration)	8.2

Napa Valley Springs Mineral Water Company

(Calistoga)

The benefits of the geysers and underground springs of Napa Valley were known to the American Indians. Then California pioneer Sam Brannan established the Calistoga Spa in 1869. The current bottling plant and warehouse are behind the stables at the old Brannan resort; the water has been bottled commercially since 1972.

The water filters through Mount St. Helena's volcanic slopes to the source which is over 200 feet deep. When it surfaces, the water is about 180° F. The spring can produce up to ten gallons a minute. Excess fluoride is removed by means of a filtration system, but otherwise the water is bottled as it comes from the source.

The water is bottled like many European waters, in one and a half liter plastic bottles, as well as in gallons. These are distributed through large supermarket chains and leading health and natural food stores.

Famous Ramona Water
Sequoia Springs Natural Pure Mineral Water

(Escondido)

The Ramona Bottling Company, located in the foothills near Ramona in San Diego County, bottles water from two different sources, producing Famous Ramona Water and Sequoia Springs Natural Pure Mineral Water. The company has bottled Famous Ramona Water since 1970, drawing the water from wells 250 and 280 feet deep.

Sequoia Springs Natural Pure Mineral Water is transported from a 1000 foot spring near the Sierra Nevada Mountains to the plant in Ramona in stainless steel tanker trucks. The water is bottled with and without added carbonation.

The water, sold in two and a half gallon plastic containers and 28 ounce bottles, is distributed through grocery stores in Los Angeles and San Diego counties. The Ramona Bottling Company also bottles their water for many private labels.

Famous Ramona Water Analysis

	Parts per Million
Calcium (Ca)	34
Magnesium (Mg)	21
Potassium (K)	3.4
Arsenic (As)	<.001
Barium (Ba)	<.5
Cadmium (Cd)	<.01
Chloride (Cl)	70
Chromium (Cr)	<.01
Copper (Cu)	.01
Fluoride (F)	.85
Lead (Pb)	<.01
Selenium (Se)	<.001
Silver (Ag)	<.01
Zinc (Zn)	.21
Bicarbonates (HCO_3)	61
Sulphates (SO_4)	42
Nitrogen (N)	.88
Nitrates (NO_3)	3.9
Phosphates (PO_4)	<.01
Phenols	.001
Total Dissolved Solids	268

Sequoia Springs Natural Pure Mineral Water Analysis

	Parts per Million
Calcium (Ca)	6.0
Magnesium (Mg)	1.2
Sodium (Na)	51.0
Potassium (K)	2.5
Aluminum (Al)	none detectable
Boron (B)	<.08
Chloride (Cl)	trace
Fluoride (F)	.3
Iron (Fe)	none detectable
Lithium (Li)	<.3
Vanadium (V)	.05
Bicarbonates (HCO_3)	155.6
Carbonates (CO_3)	trace
Nitrites (NO_2)	none detectable
Nitrates (NO_3)	none
Silica (SiO_2)	12.0
Total Hardness, as $CaCO_3$	20.0
Total Dissolved Solids	234.5
Color	5
Odor	none
Turbidity	none
pH (Hydrogen Ion Concentration)	8.6

Santé
Water Analysis

	Parts per Million
Calcium (Ca)	108
Magnesium (Mg)	34
Sodium (Na)	46
Chloride (Cl)	20
Copper (Cu)	.025
Fluoride (F)	.1
Iron (Fe)	.048
Manganese (Mn)	.022
Selenium (Se)	.005
Zinc (Zn)	.006
Bicarbonates (HCO_3)	380
Carbonates (CO_3)	1
Sulphates (SO_4)	16
Total Dissolved Hardness	556
pH (Hydrogen Ion Concentration)	6.8

Santé Water Company

(Santa Rosa)

Santé was born in northern California, the brainchild of Roland Au, a leader in the bottled water boom, from the Napa Valley area. Originally Santé's source was a geyser in Calistoga, but it discovered a better natural mineral spring in Sonoma county.

Santé was the first to introduce flavored waters to the world in 1979, and the company's success attracted Anheuser-Busch to purchase it in 1986. Aside from the carbonated mineral water, Santé also markets lime, lemon, orange, cherry, cola and root beer flavors.

Sierra Spring Water Company

(Sacramento)

Sierra Spring Water Company, founded in 1950, is the fourth largest independent producer and marketer of nonsparkling bottled water products in the United States. During its growth it has acquired 44 other bottled water companies, continuing to use many of their labels, including such locally well-known names as Sierra, Feather River Canyon, Alpine, Applegate, Blue Rock, Ampure, Airis, and Texas Spring. Under one label or another the company distributes water to 19 states and Alberta, Canada.

The company produces a variety of waters: *Drinking water*, which usually comes from the municipal water system and is treated by ozonation and filtration to remove impurities. *Spring water* comes from natural springs owned or leased by the Sierra Spring Water Company. Usually, the water is hauled from the source to a bottling plant, where it receives ozone treatment and filtration. *Distilled water* involves heating water, trapping its steam and condensing it. Distilled water contains no minerals. *Purified water*, sometimes confused with distilled water, uses the less expensive electrolysis method of removing minerals and impurities.

Silver Springs Water

(San Diego)

Silver Springs
Water Analysis

	Parts per Million
Arsenic (As)	<.005
Barium (Ba)	.01
Cadmium (Cd)	.001
Chloride (Cl)	<.001
Chromium (Cr)	<.005
Copper (Cu)	<.005
Fluoride (F)	.05
Iron (Fe)	.08
Lead (Pb)	<.005
Manganese (Mn)	<.005
Mercury (Hg)	<.001
Selenium (Se)	<.005
Silver (Ag)	<.005
Zinc (Zn)	.02
Sulphates (SO_4)	9.1
Nitrates (NO_3)	5.4
Cyanide (CN)	<.01
Phenols	.001
Total Dissolved Solids	155
Turbidity	.1

The original Silver Springs, which still supply the San Diego bottling facility, were originally used by the Cupenio Indian tribe of the Dieguito nation, who often camped at Silver Springs near the mountain community of Ramona, California. Relics are still being found there. After these Indians moved on, the Silver Springs were partially covered over and forgotten. In 1930, rancher George Smith decided to build a grape arbor in this area. In digging postholes, Smith noticed bubbling water; he was impressed by its flavor and had it analyzed. The analysis showed a water astonishingly free from sediment or bacteria. George Smith subsequently bottled the water in limited quantities, filling gallon bottles from a hand operated pitcher pump and marketing them door to door in Coronado, California.

Today the company has bottling plants in San Diego and Orange, as well as distribution facilities in Gardena and San Marcos, California. Water sold from the Orange County bottling plant is from a deep well. The water bottled at both bottling facilities is blended with distilled well waters to produce a pleasant tasting, mild mineral water. The company currently sells in excess of ten million gallons of Silver Springs Water annually.

Silver Springs Water services San Diego County, Orange County and the greater Los Angeles-Riverside County areas. The principal packages are the traditional five gallon bottle and a three gallon case of half gallon bottles.

Sparkletts Drinking Water Corporation

(Los Angeles)

Sparkletts Crystal-Fresh Water Analysis

	Parts per Million
Calcium (Ca)	.7
Magnesium (Mg)	.8
Sodium (Na)	3.8
Chloride (Cl)	3.5
Fluoride (F)	0.0
Bicarbonates (HCO_3)	5.8
Carbonates (CO_3)	0.0
Sulphates (SO_4)	3.4
Nitrates (NO_3)	0.0
Total Hardness, as $CaCO_3$	5.1
ph (Hydrogen Ion Concentration)	6.8

The Sparkletts Drinking Water Corporation was organized in 1925 as the Sparkling Artesian Water Company, utilizing wells located at 4500 York Boulevard in the Eagle Rock section of Los Angeles. Its name was later changed to the present one. In 1964, Foremost Dairies purchased Sparkletts. The later merger of Foremost with McKesson-Robins has helped Sparkletts grow into the largest bottling water company in the country, and the key operating unit of McKesson Water Division.

Well-water is prepared by carbon sand filtration and ozonation reverse osmosis and/or deionization to make Purified (mineral-free) water. Selected minerals are added to Purified water, by formula, to prepare Crystal-Fresh drinking water. The company also bottles Crystal-Fluoridated (Crystal-Fresh with fluoride added).

Crystal-Fresh, Purified and Crystal-Fluoridated waters are delivered to homes and offices in five-gallon bottles. Crystal-Fresh and Purified are available in gallon and two and one-half gallon containers, and at retail outlets from Santa Barbara to the Mexican border and as far east as Las Vegas.

Vichy Springs

(Ukiah)

The 100,000-year-old Vichy Springs Grotto is located in Ukiah, Mendocino County. The Pomo Indians are known to have used the grotto as a holy site and the mineral water as a medicinal tonic as early as 1000 B.C. In 1854, Mr. Day began a resort he called Day's Springs; then, in 1865, a Col. Doolan bought the springs, renaming them after the famous European springs, and expanded the resort to include, among other attractions, an early version "jacuzzi"; its famous "Champagne Bath," a highly effervescent, sparkling water, flowing from the spring at a temperature of 90 degrees. The springs have known four owners, but their use as a resort and a source of bottled water has remained constant.

World travelers Gilbert and Marjorie Ashoff tested and tasted the waters of 150 California springs before forming the Vichy Springs Mineral Water Corporation to purchase the springs in 1977, captured by the flavor, the laboratory test results and the ambience of Vichy Springs, California.

Nine separate springs, all different, from hot to cold, carbonated to noncarbonated, are found on the property. The corporation invested in a new label, a heavyweight six-pack carrier, to market its sparkling mineral waters that have been filtered to remove oxidized iron content and have additional CO_2 to offset losses during the filtration process.

Vichy Springs Water Analysis

	Parts per Million
Calcium (Ca)	157
Magnesium (Mg)	48
Manganese (Mn)	.02
Sodium (Na)	1095
Potassium (K)	38
Chloride (Cl)	262
Copper (Cu)	.02
Fluoride (F)	.36
Lithium (Li)	1.05
Strontium (Sr)	2.3
Bicarbonates (HCO_3)	3210
Nitrates (NO_3)	.09
Sulphates (SO_4)	.1
Total Hardness, as $CaCO_3$	590
Total Dissolved Solids	3170
pH (Hydrogen Ion Concentration)	6.8

Vittel/Bartlett Water Analysis

	Parts per Million
Calcium (Ca)	67
Magnesium (Mg)	360
Sodium (Na)	4.3
Potassium (K)	.74
Chloride (Cl)	8
Iron (Fe)	.03
Sulphates (SO_4)	21.5
Bicarbonates (HCO_3)	1995
Nitrates (NO_3)	0

Vittel/Bartlett Springs

(Bartlett Springs)

At the turn of the century Bartlett Springs, a thriving resort spa, lured guests by the thousands up a difficult road for posh weekends of water drinking and mountain air. "One of the greatest natural sanitariums in the world," with its first class hotels, steam and mineral baths attended by therapists and masseuses, a ballroom, casino and concert hall, Bartlett Springs had a world-wide reputation that attracted European aristocracy, including Queen Marie of Rumania.

Its underground mineral springs had been named for the man who discovered them in 1870, Green Bartlett. Bartlett credited the spring with stopping his rheumatic misery, filed a claim on it, together with his partner, L. Tharp, built cabins for fellow sufferers and began bottling water.

In 1983, after a five-year search for an American spring that could match the quality of its European spring, the Vittel Mineral Water Company of France coventured with Nestle and purchased Bartlett Springs for one million dollars. Vittel, too, is calling attention to the health-promoting qualities of Vittel/Bartlett's magnesium-calcium-rich water. In addition to its nonflavored, naturally sparkling mineral water, Vittel/Bartlett's is also available in lime, lemon-lime, cherry and orange flavors.

COLORADO

The original Manitou Mineral Water Company label, c. 1900.

Manitou Mineral Water

(Manitou Springs)

Seventy-eight miles south of Denver is one of the most famous springs in the United States, Manitou Springs. Its written history dates back to the Ute Indians who gave it the name Manitou, used to describe occult manifestations of nature such as waterfalls and springs. (The Arapahos called it Medicine Fountain. They believed a spirit lived in the spring; when he exhaled, bubbles came to the surface.) The Indians would leave offerings for the spirit of the spring, consisting of blankets, bows, beads, deerskin, knives and moccasins thrown in the water or hung in trees growing nearby.

In 1806 Lieutenant Zebulon Pike discovered the mountain peak which bears his name, but he missed the bubbling springs on the mountainside. It was Dr. Edwin James, a historian accompanying the Long expedition, who visited Pike's Peak region in 1820 and wrote:

> "The boiling spring is a large and beautiful fountain of water, cool and transparent, and aerated with carbonic acid. It rises on the brink of a small stream which here descends from the mountains at the point where the bed of this stream divides the ridge of sandstone, which rests against the base of the first granite range. The water of the spring deposits a copious concretion of carbonate of lime, which has accumulated on every side, until it has formed a large basin overhanging the stream, above which it rises several feet. The basin is of snowy whiteness and large enough to contain 300 or 400 gallons, and is constantly overflowing. The spring rises from the bottom of the basin with a rumbling noise, discharging about equal volumes of air and water, probably about fifty gallons per minute, the whole kept

in constant agitation. The water is beautifully transparent, has a sparkling appearance, the grateful taste and exhilarating effect of the most highly aerated artificial mineral water."

The year 1886 marked the first commercial bottling of Manitou spring water. The Ute Chief Mineral Water Company and Hiawatha Bottling Works (as the company was then known) also introduced bottled sarsaparilla and ginger champagne. A new bottling plant was built in 1891 and the company was sold to Mr. F. J. Schueler, co-founder with Adolph Coors of the Coors Brewing Company. Manitou Spring Water was bottled by the Schueler family until the 1950s. The Manitou Corporation re-introduced this excellent mineral water, which has a faint sweet flavor, to the public in the fall of 1978.

A view of the Manitou Soda Springs around 1890.

FLORIDA

Good Hope Water Company

(Jacksonville)

Good Hope Water Company was established in 1911 when the spring's owner decided to bottle the water. Previously the water had been given to those who journeyed to the spring with empty bottles. Until 1971 it remained a family owned business, when it was purchased by the Leasing Corporation.

Jacksonville has had various problems with its municipal water source. Once red worms were found in the water. More recently traces of ocean water have been found in the municipal supply, which indicates the sea is being drawn into freshwater supplies due to dropping tables. A recent proposal has been made to limit the digging of wells in the area.

Good Hope Water comes from a well source 900 feet deep. It is piped through a charcoal filter and labeled Good Hope Mineral Water. The Purified brand is further processed by reverse osmosis and deionization. Both waters are available in one and five gallon glass bottles.

Good Hope Water Analysis

	Parts per Million
Calcium (Ca)	59.68
Magnesium (Mg)	21.04
Sodium (Na)	10.44
Chloride (Cl)	18
Fluoride (F)	.72
Iron (Fe)	.05
Bicarbonates (HCO_3)	125
Sulphates (SO_4)	97
Total Hardness as $CACO_3$	250
Total Dissolved Solids	373

Zephyrhills Bottled Water

(Zephyrhills)

Zephyrhills
Bottled Waters
Water Analysis

	Parts per Million
Calcium (Ca)	42
Magnesium (Mg)	2.2
Sodium (Na)	0
Chloride (Cl)	5.9
Fluoride (F)	1
Iron (Fe)	.08
Bicarbonates (HCO_3)	35
Sulphates (SO_4)	5
Total Hardness as $CaCO_3$	114
Total Dissolved Solids	150

The town of Zephyrhills, located in the quiet countryside of central Florida, has long been called the "City of Pure Water," and is also known for one of the best-tasting waters in Florida. The Zephyrhills Bottled Water Company was formed in the early 1960s and is currently owned by The Perrier Group. The products include natural, purified, and distilled water in a variety of package sizes for both grocery stores, and for home and office delivery.

Zephyrhills Bottled Water comes from a more than thousand-foot-deep underground spring, the deepest in the state of Florida, and is surrounded by more than 400 wooded acres. The water is filtered through many layers of sand and limestone.

HAWAII

Pure Hawaiian Sparkling Water

(Honolulu)

The Hawaiian island of Oahu possesses a fascinating geologic phenomenon called a Guyben-Hertzberg lens. It was first discovered in Hawaii, around 1911, that rainwater infiltrates the island's volcanic rock and forms a convex lens of fresh water between the rock and the seawater. This lens traps pure rain water which provides much of Oahu's drinking water and was the original source for Pure Hawaiian Sparkling Water when it was introduced in 1983.

However, Marcus Bender, founder and president of Hawaiian Water Partners, discovered an even purer source on the Big Island in Puna between the Mauna Loa and Mauna Kea volcanoes, situated on 16,000 protected acres of land. Rain water filters naturally through volcanic rock hundreds of feet below the surface where it is collected for bottling in Hawaii and on the mainland. The water has a remarkably low natural total dissolved solids of 54 milligrams per liter.

Pure Hawaiian Sparkling Water is the first domestically produced natural water packaged in an aluminum can. It is also packaged in 10-ounce glass and a one-liter PET bottle. Some of the water is shipped in bulk, for packaging on the mainland, in 5,300-gallon flexitanks developed by Hawaiian Water Partners. Pure Hawaiian Sparkling Water is also available in six flavors: guava-cherry, papaya-coconut, passion-orange, lime, lemon and orange.

Pure Hawaiian Sparkling Water Water Analysis

	Parts per Million
Calcium (Ca)	5
Magnesium (Mg)	3
Sodium (Na)	10
Potassium (K)	25
Chloride (Cl)	10
Fluoride (F)	.5
Iron (Fe)	.1
Hardness (as $CaCO_3$)	19
Total Dissolved Solids	54

ILLINOIS

Hinckley & Schmitt Water Company

(Chicago)

Hinckley & Schmitt is a company that grew out of a plague! In 1888, typhus was swarming in Chicago's drinking water. George J. Schmitt, a pharmacist assistant, met Otis Hinckley, a salesman from Wisconsin, selling spring water to customers, many of whom had prescriptions for it from their doctors. It was the beginning of a company that eventually bottled and distributed waters from six mineral springs in Wisconsin and Illinois — White Rock Mineral Springs, Corinnis Waukesha Springs, Silurian Mineral Springs, Waukesha Arcadian Springs, Aurora Lithia Springs and Blue Grass Springs. Recently the company purchased the "5 Springs" Water Company in El Paso, Texas.

Hinckley & Schmitt was the largest independent water company in the United States until 1981, when the firm was purchased by Anjou International Company, a wholly owned subsidiary of France's Compagnie Generale des Eaux (CGE), one of France's largest publicly traded utilities. With a slogan emphasizing the body's need for eight glasses of water a day — Eight is Great! — the company is now marketing bottled water in 13 states.

LOUISIANA

Kentwood Spring Water

(New Orleans)

From a city known to have some of the worst *municipal* water in the United States comes the Kentwood Bottling Company. The water source, an artesian well 1,000 feet deep, is located in the town of Kentwood 100 miles north of New Orleans. Since 1965 the company has trucked water from the source to an immaculate bottling plant in New Orleans. Kentwood Spring Water is so concerned with purity that even air entering the bottling room is filtered and sterilized.

The water is available in one gallon plastic bottles as well as in half gallon glass bottles. The company also produces distilled water and fluoridated water.

In 1986, Kentwood became part of Suntory International of Japan. Currently, the company produces various water products, including Artesian Spring, Fluoridated, Distilled and Sparkling Artesian. The water is available in five-gallon, two and one half gallon, one-gallon, one-half gallon, one-liter and ten-ounce bottles.

Kentwood Spring Water Analysis

	Parts per Million
Calcium (Ca)	2.8
Magnesium (Mg)	.5
Sodium (Na)	7.4
Chloride (Cl)	3.4
Bicarbonates (HCO_3)	21.4
Sulphates (SO_4)	2.9

MAINE

Ice Mountain Spring Water Analysis

	Parts per Million
Calcium (Ca)	1.2
Magnesium (Mg)	.8
Sodium (Na)	.5
Potassium (K)	.4
Chloride (Cl)	.6
Silica (Si)	24.8
Bicarbonates (HCO_3)	12
Total Dissolved Solids	20
pH (Hydrogen Ion Concentration)	6.27

Ice Mountain Spring Water

(Mt. Zircon)

Mt. Zircon Spring is located on the 3,300 acres of protected forest surrounding Mt. Zircon. It was first introduced to the public in 1859, and became renowned for its curative properties. Subsequently, the water was sold under the Moontide label until it was purchased in December of 1986, by Pepsi Company's Great Spring Waters of America division.

Ice Mountain is available in one, two and one half, three and five-gallon bottles. It is distributed through supermarts and by home delivery.

Poland Water Analysis

	Parts per Million
Calcium (Ca)	18.5
Magnesium (Mg)	7.2
Sodium (Na)	4.75
Potassium (K)	.65
Chloride (Cl)	2.8
Iron (Fe)	<.05
Bicarbonates (HCO_3)	78.0
Sulphates (SO_4)	7
Nitrates (NO_3)	13.6
Silica (SiO_2)	20.3
Total Hardness, as $CaCO_3$	57.8
Total Dissolved Solids	124
pH (Hydrogen Ion Concentration)	7.27

Poland Spring

(South Poland)

The oldest American bottled natural spring water, Poland Spring Water, has been sold commercially since 1793.

In the nineteenth century, health-minded guests from the east coast gathered in Poland Spring to take the waters. Then, in 1845, Hiram Ricker bottled the water, enabling visitors to take it home with them. Ships captains and stage drivers stocked up on it for their trips.

Poland Spring still comes from one deep underground spring in Poland Spring, Maine. The granite spring is protected by over 400 acres of wilderness in the Maine backwoods. The water surfaces crisp and cold, the result of a glacier that covered Maine 10,000 years ago.

Poland Spring delivers its bottled water to homes and businesses throughout New England and in the New York metro area. It is also available in grocery stores as nonsparkling water, a sparkling variety and a line of sparkling natural essence flavors. In 1988, Poland Spring introduced "sparkling juice," which blends its own spring water with fruit juice.

Poland Spring is a company of The Perrier Group.

Den Haene found a natural source of gas in Colorado, and began producing a carbonated version of Poland Spring. The Perrier Group later repurchased Poland Water. In 1988, Poland began marketing blended natural juice and spring water, following the example of its sister company, Calistoga.

MASSACHUSETTS

Belmont Springs Water Company

(Belmont)

Belmont Springs was a popular Sunday excursion for Bostonians a century ago. The owner, George Cotton, supplied a silver dipper for his visitors' use. In 1922, John McGory bought the company. Four horse and wagon teams delivered five gallon jugs for 50 cents

Belmont Springs
Water Analysis

	Parts per Million
Calcium (Ca)	26
Magnesium (Mg)	4.3
Sodium (Na)	3.3
Potassium (K)	.65
Aluminum (Al)	.07
Barium (Ba)	.06
Boron (B)	.01
Chloride (Cl)	23
Chromium (Cr)	.002
Copper (Cu)	.01
Iron (Fe)	.03
Lead (Pb)	.012
Manganese (Mn)	.003
Molybdenum	.006
Nickel	.007
Phosphorous	.023
Silver (Ag)	.01
Strontium (Sr)	.13
Sulphates (SO_4)	44.2
Nitrites (NO_2)	.03
Nitrates (NO_3)	.34
Total Hardness, as $CaCO_3$	120
Total Dissolved Solids	188
pH (Hydrogen Ion Concentration)	7.63

throughout Boston. The water was even canned and used in life boats during World War I. The company was sold to Coca Cola in 1969 and now bottles about 13 million gallons annually. Bostonians not only drink this spring water but fill their goldfish bowls with it, and local swimming pools have even been filled with Belmont Springs Water.

MINNESOTA

Glenwood-Inglewood Spring Water

(Minneapolis)

Glenwood-Inglewood Spring Water Analysis

	Parts per Million
Calcium (Ca)	95
Magnesium (Mg)	38
Sodium (Na)	15
Potassium (K)	7
Aluminum (Al)	0
Chloride (Cl)	12
Copper (Cu)	0
Fluoride (F)	.34
Iron (Fe)	0
Lead (Pb)	0
Manganese (Mn)	0
Mercury (Hg)	0
Silver (Ag)	0
Zinc (Zn)	0
Sulphates (SO_4)	86
Nitrites (NO_2)	0
Nitrates (NO_3)	0
Phosphates (PO_4)	.02
Silica (SiO_2)	28
Surfactants (Detergent)	0
pH (Hydrogen Ion Concentration)	7.3

In the late nineteenth century a young Englishman, seeking his fortune, established a hardware manufacturing company in Minneapolis. Searching for drinking water for his workmen, he found a beautiful, clear spring near his factory. Unknowingly, W. H. Fruen had laid the foundation for one of the leading bottled water companies of the present day.

Glenwood Spring Water Company first began to bottle water in 1884. In 1896 the second generation of Fruens took charge of the company. They arranged to merge with a competitor, Inglewood Spring Water Company, forming the Glenwood-Inglewood Company. In 1950, grandson J. Donald Fruen assumed presidency, and in 1972, Todd C. Fruen assumed command, making the fourth generation in the business.

An estimated 10,000 gallons of water flows through the spring each hour, Glenwood-Inglewood bottles spring and distilled water in various size bottles from one-half gallon to six-gallon.

MONTANA

Big Springs Bottled Water

(Lewistown)

From 4,000 feet high in the Snowy Mountains of Montana flows one of the largest natural springs in the United States. The spring provides the 60,000 inhabitants of Lewistown with fresh unchlorinated drinking water. It surfaces at a constant 52° F, at the staggering rate of 90 million gallons a day.

Big Springs Water is bottled by Coca Coca Bottling Company and distributed to grocery stores around Lewistown and to surrounding states. The water is available only in one gallon jugs.

Big Springs Bottled Water Analysis

	Parts per Million
Calcium (Ca)	70
Magnesium (Mg)	16
Sodium (Na)	2
Potassium (K)	2
Arsenic (As)	0
Chloride (Cl)	4
Copper (Cu)	0
Fluoride (F)	.4
Iron (Fe)	0
Lead (Pb)	0
Bicarbonates (HCO_3)	191
Carbonates (CO_3)	0
Sulphates (SO_4)	77
Nitrates (NO_3)	.5

NEW MEXICO

Springsweet Deep Well Water

(Albuquerque)

In the 1930s the well which now provides water for Springsweet was drilled at an altitude of 4,945 feet in the Rio Grande Basin near Albuquerque. The well is 220 feet deep. The water is filtered through charcoal and then a paper depth filter, and piped through an ultraviolet sterilizer. Available only in five gallon glass bottles, the water is distributed throughout New Mexico.

Springsweet Deep Well Water Analysis

	Parts per Million
Calcium (Ca)	.05
Sodium (Na)	41.4
Barium (Ba)	1
Cadmium (Cd)	.01
Chloride (Cl)	30.5
Chromium (Cr)	.05
Copper (Cu)	.01
Fluoride (F)	.89
Iron (Fe)	.3
Lead (Pb)	.05
Manganese (Mn)	.05
Selenium (Se)	.01
Silver (Ag)	.05
Zinc (Zn)	.05
Bicarbonates (HCO_3)	141
Sulphates (SO_4)	47
Nitrates (NO_3)	1.8
Cyanide (CN)	.01
Total Hardness, as $CaCO_3$	37
Total Dissolved Solids	328

NEW YORK

Deer Park 100% Spring Water

(New York)

Near the crest of the scenic Alleghenies, in Maryland, the Deer Park summer resort and hotel was built in 1873. It had the advantage of fresh air, unmatched scenic vistas and pure mountain spring water unsurpassed anywhere in the country. It quickly became a favorite spa for America's vacationing elite, including at least four presidents: Cleveland, Harrison, Garfield and Taft. Servicing the fashionable resort was the B&O Railroad which served Deer Park spring water with its "delightful palatable taste" to patrons in the plush dining cars.

The original clear mountain spring water continues to serve as a

Deer Park 100% Spring Water Analysis

	Parts per Million
Magnesium (Ma)	3
Sodium (Na)	1.7
Arsenic (As)	<.01
Barium (Ba)	.02
Cadmium (Cd)	<.01
Chloride (Cl)	10
Chromium (Cr)	<.01
Copper (Cu)	<.01
Fluoride (F)	<.05
Iron (Fe)	<.01
Lead (Pb)	<.01
Manganese (Mn)	<.05
Selenium (Se)	<.01
Silver (Ag)	<.01
Zinc (Zn)	<.01
Sulphates (SO_4)	4
Nitrates, as N	.05

standard for quality and flavor for today's Deer Park 100% Spring Water. Now, as at the turn of the century, no purifiers, such as chlorine, are added to Deer Park water. Convenient half gallon and gallon bottles are available in supermarkets and grocery stores.

Phenols	<.001
Total Hardness, as $CaCO_3$	10
Alkalinity	14
Total Dissolved Solids	29
Color	<5
Odor	None
Turbidity	.42
pH (Hydrogen Ion Concentration)	6.06

Great Bear Spring Water Company
(New York)

As one of the oldest American bottled water companies, Great Bear has been quenching the thirst of the Northeast with its still water since 1888, when the company was founded. Great Bear operates ten branch locations with delivery routes in 12 states. Its line of water products is also available through grocery stores.

Recently the company was purchased by The Perrier Group.

Saratoga Springs
(Saratoga)

Rumor has it that the springs were discovered by Iroquois Indians, back before the white man arrived. Later, sipper boys at the spa ladled Saratoga Springs water into monogrammed cups for Lillian Russell and her beau, Diamond Jim Brady, during the Gay Nineties. Big draughts of it were gulped by Secretariat when he was stabled at the thoroughbred track in 1973. Other visitors to Saratoga Springs have included Sir William Johnson, General Schuyler, Alexander Hamilton, George Washington, and Gideon Putnam.

The springs are of two kinds, saline, which are laxative, and alkaline, which are settling to the digestive tract. The waters are naturally carbonated—they contain carbonic acid when they flow from the earth. Some of the wells have been producing continuously for over a century with little or no change in chemical composition.

At one time or another there were approximately one hundred named mineral springs in Saratoga, about equally divided between the city proper and the lands of the present New York State Spa Park.

Home delivery of legendary Saratoga Springs mineral water. Wagon used as late as 1930.

Of these, six remain in the city, available free to the public.

Source: The mineral waters all originate in a layer of dolomitic limestone that underlies the shale and glacial drift of the area. The original springs bubbled through to the surface through faults in the shale.

Radioactivity: All natural waters are radioactive to some extent. Deep seated waters usually contain a greater concentration of radioactive elements than surface waters. The mineral waters of the Saratoga Springs Reservation are of the deep seated type. Radium 226 is the most important of the naturally occurring radioisotopes in water because it has a tendency to accumulate in the bone structure of the body.

Supply: Supply seems to be limited, in that the wells must be allowed to replenish themselves naturally. Over-use of the wells could possibly lead to their extinction. It was this possibility which caused the State of New York to assume control over the mineral water rights. Eighteen mineral wells are currently producing on the Saratoga Springs Reservation. They are Hathorn No. 3, Coesa, Geyser, Orenda, Hayes, Island Spouter, Karista, Polaris, Pump Well

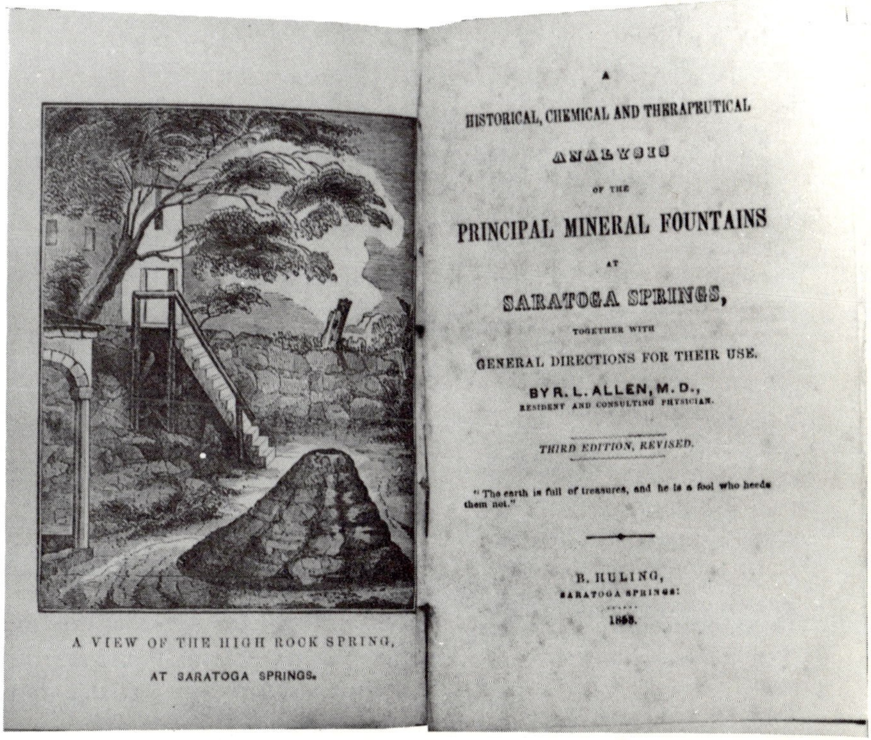

A VIEW OF THE HIGH ROCK SPRING,
AT SARATOGA SPRINGS.

A
HISTORICAL, CHEMICAL AND THERAPEUTICAL
ANALYSIS
OF THE
PRINCIPAL MINERAL FOUNTAINS
AT
SARATOGA SPRINGS,
TOGETHER WITH
GENERAL DIRECTIONS FOR THEIR USE.
BY R. L. ALLEN, M. D.,
RESIDENT AND CONSULTING PHYSICIAN.
THIRD EDITION, REVISED.
"The earth is full of treasures, and he is a fool who heeds them not."
B. HULING,
SARATOGA SPRINGS:
1853.

The title page of R.L. Allen's 1853 publication on Saratoga Springs. High Rock Spring, pictured on the right, is considered to be the primeval fountain. Centuries of calcareous deposits formed this remarkable monument to mineral waters.

No. 5, Champion No. 2, and eight Lincoln wells which are combined and used for the mineral water baths.

The original spring was hand-drilled through 30 feet of sand and 150 feet of rock in 1872. Its sparkling mineral water was hand-bottled, with cork-and-wire stoppers, and distributed by horse and wagon throughout New York. In 1985, Anheuser-Busch Beverage Group purchased Saratoga Naturally Sparkling Mineral Water, with New York as its principal market and a 15 percent annual increase in its growth rate predicted through 1990.

The effervescence comes from natural carbon dioxide taken from the same underground source as the mineral water itself.

Ferndell Spring is a fresh water source available at an outlet in Geyser Park. This water is different from the mineral waters in that it is much lower in mineral content, and does not contain the excess carbon dioxide that causes the mineral waters to effervesce.

PENNSYLVANIA

Cloister Pure-Spring Water Company

(Ephrata)

In the early 1890s Professor Leisening, an expert on water cures, established a hydropathic institute in Ephrata, due to the mineral analysis of the town's famous mountain spring water. Directly across from the Institute was the Hotel Cocalico, which had its own supply of spring water, then known as Glacier Springs. It was piped into the hotel and was available to the guests. For $1.50 a day, one could buy a room, meals and water.

Ephrata's spring water came to the surface through red sandstone rocks, impregnated with iron and magnesia. The water

Cloister Pure-Spring Water Analysis

	Parts per Million
Sodium (Na)	3.1
Potassium (K)	1.5
Aluminum (Al)	.14
Chloride (Cl)	4
Fluoride (F)	nil
Iron (Fe)	nil
Manganese (Mn)	nil
Sulphates (SO_4)	1
Nitrites (NO_2)	nil
Nitrates (NO_3)	.25
Phosphates (PO_4)	.2
Ammonium (NH_4)	nil
Surfactants (Detergent)	nil
Total Hardness, as $CaCO_3$ in grains per gallon	1.5
Total Dissolved Solids	42
Color	nil
Odor	nil
Turbidity	nil
pH (Hydrogen Ion Concentration)	5.7

is soft, alkaline mineral spring water, free of chlorine and other additives. Early newspapers describe the "therapeutic value of the Springs at the Old Cocalico Hotel on Ephrata Mountain." The water was said to be helpful for kidney disorders and excellent for the removal of uric acid. It was believed this water would positively prevent typhoid fever and other germ-caused diseases. It was even once believed that the water would prevent hair from falling out, and keep the scalp free of dandruff.

No such marvelous claims are made today. However, the water is delicious, sparkling and clear. This natural spring water is now being bottled by Cloister Pure-Spring Water Company in half gallon, one gallon and five gallon bottles.

Ephrata Diamond Spring Water

(Ephrata)

From its beginnings as a community of German pietists seeking to build a life of purity, culture and industry, the town of Ephrata owed its existence to its water that is naturally filtered through a ridge of sandstone, unique among the limestone foundations of most of Lancaster County. Located in the heart of the Pennsylvania Dutch section of the state, Ephrata grew into a widely known spa.

Diamond Spring was discovered by a Mr. Herr in 1895. Recognizing its superior taste, he began "bottling" it in milk cans, and selling it on the streets of Lancaster, pouring the Diamond Spring water into his customers' own containers and taking home his empties to be refilled.

The present owner, Richard Hess, controls three separate properties with more than 100 acres of mountain land containing springs and wells. The water is pumped slowly into glass-lined tanks, and transported to a plant, where it is filtered and treated by the Hesone ozone sterilization system.

Diamond Springs offers noncarbonated spring and distilled water in gallon plastic containers.

Diamond Spring Waters

Water Analysis

	Parts per Million
Sodium (Na)	10
Chloride (Cl)	4
Fluoride (F)	.1
Iron (Fe)	.02
Manganese (Mn)	.01
Zinc (Z)	.01
Sulphates (SO_4)	5
Total Hardness as $CaCO_3$	66
Total Solids	184
pH (Hydrogen Ion Concentration)	7

Polar Water Company

(Pittsburgh)

Polar Water Company was founded in Pittsburgh in 1900, by J.M. Barrett. The company was incorporated in 1905, and in 1972, it was purchased by Borden, Inc.

Both Polar Water Company and Crystal Water Company were purchased by Beverage Management Inc., in December 1978.*

Currently, the Polar Water Company has a manufacturing and bottling plant in Pittsburgh, Pennsylvania, from which radiate its wholesale and retail operations. Producing both "Distilled" water and "Drinking" water packaged in one (and more) gallon containers, the company markets its product in Pennsylvania, West Virginia, Ohio, Maryland and Washington, D.C.

Crystal Water is a Florida-based operation with facilities from Cocoa Beach to the Florida Keys.

Polar and Crystal market four kinds of water in 22 packages: drinking water, natural water, spring water, and distilled water. Their waters are largely sold in the five-gallon containers.

*Beverage Management's Water Division was purchased by Suntory International in December 1987.

Crystal Clear Drinking Water Water Analysis

	Parts per Million
Iron (Fe)	.02
Sulphates (SO_4)	2
Total Hardness (as $CaCO_3$)	40
Total Solids	76
pH (Hydrogen Ion Concentration)	7

Short Mountain Spring Water Company

(Short Mountain)

A newly-formed company dedicated to providing healthful drinking water, its slogan, "Kick the tap habit!" parallels the company's attitude that "when less than one percent of the world's total supply of water is suitable for human consumption and two-thirds of our body is made of water," extraordinary measures are called for. Short Mountain Spring Company bases its promotion on the purity of its water, "one of the finest ever tested."

Valley Forge Springs

(Norstown)

In 1914 Morris Bore discovered a spring while he was excavating for a house. Subsequent findings of tools and implements proved that Indians had lived in this area. Bore then located three other springs on his property, one with considerable iron, another with a considerable amount of lime, and another that was alkaline. The first one he found was said to be "plain water."

In his exploration for water Bore also found a vein of tin ore, and he mined various gems such as milk opal, topaz, garnet, neck blue sapphire and moonstone, having uncovered the gems in a sand dune while planting potatoes.

For many years the mineral waters were bottled and people said they relieved indigestion, rheumatism, bronchial catarrh, sluggish liver, swollen glands, dizziness, gastroenteritis, heartburn and other afflictions. Bore made his first deliveries on foot and later deliveries in automobiles and Model A trucks.

Although the mineral springs have dried up, the "plain water" is still bottled, and provides jobs for four driving salesmen.

Wissahickon Spring Water Company

(Manheim)

The company began in the determination of Maurice J. Land, father of the current owners, to provide pure, tasteful, odorless drinking water. In 1970, he found and purchased springs that met his standards in Manheim, a small town in the hills of Lancaster County. The spring's ability to please the palate had already been demonstrated, having been used commercially since the 1800s, when their waters were transported by horse and wagon to the railroad station and shipped in tank cars to New York distributors. A modern bottling plant was quickly put into operation, and seven years later, demand for Wissahickon Spring Water required construction of a second plant, bottling in one-gallon and half-gallon plastic containers.

Under the management of James J. Land and Joseph J. Land, the company is serving a rapidly growing market spanning the East coast.

Wissahickon Spring Water Water Analysis

	Parts per Million
Arsenic (As)	0.01
Barium (Ba)	0.1
Cadmium (Cd)	0.001
Chloride (Cl)	0.5
Chromium (Cr)	0.001
Copper (Cu)	0.001
Fluoride (F)	0.04
Iron (Fe)	0.01
Manganese (Mn)	0.028
Nitrogen, Nitrate (N)	0.1
Selenium (Se)	0.01
Silver (Ag)	0.01
Zinc	0.01
Sulphates (SO_4)	18
Total Dissolved Solids	41

TEXAS

Artesia Water, Inc.

(San Antonio)

Artesia Water Analysis

	Parts per Million
Calcium (Ca)	19.9
Magnesium (Mg)	2.8
Chloride (Cl)	6.4
Fluoride (F)	.06
Iron (Fe)	.007
Bicarbonates (HCO_3)	59
Nitrates (NO_3)	1.3
Sulphates (SO_4)	10
Total Alkalinity as $CaCO_3$	48.3
Total Dissolved Solids	73.9

Less than a decade old, Artesia Water, Inc., draws its water from the Edwards aquifer, a 170-mile-wide and 50-mile-long underground "lake" near San Antonio, Texas.

The sparkling water, which has a rich natural mineral content, has ozone and carbonation added.

Rick Scoville, president and originator of the company, has impressed the bottled water industry with the success of his advertising program. Designing a bottle that looks suspiciously like a beer bottle, Scoville backed it with a macho campaign that appealed to Texan pride, featuring the slogan: "Artesia — Pure Texas Spirit. A sparkling mineral water from Texas Hill Country. *Au Revoir*, Perrier." And then, as his marketing went national, so did his slogan: "Pure American Spirit — *Au Revoir*, Perrier and Evian!"

In 1985, the company added a new product, Artesia — "A star-spangled 'Stil$_{TM}$' — but marketed in the same amber bottle under a new label.

Recently, Artesia has developed a line of flavored water containing no sugar, sodium or salt.

Oasis

(Camp Tahuaya)

For hundreds of years, the Oasis Spring has been renowned for the purity and taste of its water. Cowboys, homesteaders, and even the legendary Butch Cassidy are said to have taken the long, grueling ride through the Texas hills just to drink the cool, crystal-clear water of the spring. With the slogan, "One of Texas' Great Natural Resources," the company holds a prominent position in the Texas bottled water market. Selling three types of water — pure spring, drinking and distilled — Oasis is available largely through food stores.

Oasis Spring comes from a protected source in Bell County, Texas, that yields a good-tasting, naturally salt-free water.

The company has recently been acquired by The Perrier Group.

Ozarka Water Company

(San Antonio)

The Tula Indians, who fought against De Soto and his men at Caddo Valley in 1541, drew their water from a spring on the west side of the Oachita Mountains in the Ozarks of Arkansas.

The Ozarka Water Company collects the water from this spring in enormous stainless steel tank trucks and transports it to the processing plant in San Antonio and other franchised outlets in Texas, Oklahoma and Kansas, where it is bottled under the label of Caddo Valley Ozark Mountain Spring Water.

The water from this historic spring surfaces at a constant 59° F at the rate of 200,000 gallons per day.

Ozarka has been purchased by The Perrier Group.

Ozarka Water Co. Water Analysis

	Parts per Million
Calcium (Ca)	41.6
Magnesium (Mg)	.49
Sodium (Na)	trace
Chloride (Cl)	1.02
Fluoride (F)	.22
Iron (Fe)	.05
Manganese (Mn)	0
Bicarbonates (HCO_3)	119.56
Carbonates (CO_3)	0
Sulphates (SO_4)	2.02
Silica (SiO_2)	8.50
Free CO_2 gas	17
Total Hardness, as $CaCO_3$	106
Total Dissolved Solids	123
pH (Hydrogen Ion Concentration)	7.1

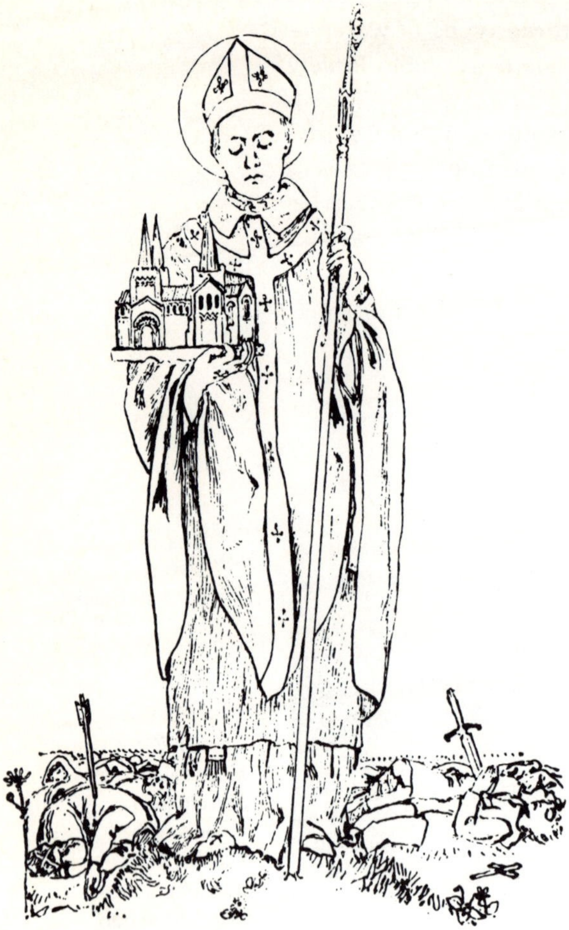

Saint Chad, patron saint of medicinal springs and wells.

UTAH

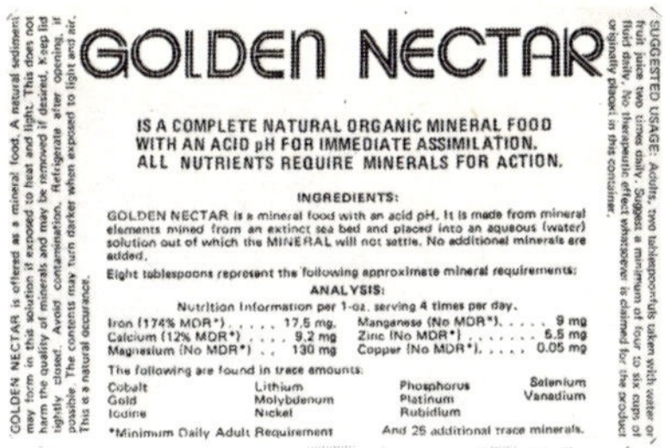

Golden Nectar
(Salt Lake City)

Golden Nectar "organic mineral food" is known for its therapeutic effects. This unusual combination of minerals, which the Indians prized highly for its healing qualities, is mined from a dry sea bed in a remote mountain valley in Utah and put into a suspended solution for distribution around the country.

Although this highly mineralized water has an unpleasant taste, natural healers have reported success in treating such varied ailments as bad eyesight, sore throat, arthritis, hemorrhoids, ulcer, burns, scar tissue and gall bladder complaints.

An analysis performed in 1976 showed the extremely high mineral content of Golden Nectar. According to the laboratory, the exact distribution of minerals is quite changeable, so this analysis is only representative.

Golden Nectar
"Liquid Mineral #2"
Water Analysis

	Parts per Million
Calcium (Ca)	80
Magnesium (Mg)	850
Sodium (Na)	170
Potassium (K)	200
Aluminum (Al)	2,160
Antimony	10
Arsenic (As)	20
Barium (Ba)	10
Bismuth	10
Boron (B)	10
Cadmium (Cd)	10
Carbon (C)	10
Cesium	10
Chloride (Cl)	10
Chromium (Cr)	10
Cobalt	10
Copper (Cu)	80
Fluorine (F)	30
Gold	10
Iodine (I)	10
Iron (Fe)	80
Lead (Pb)	10
Lithium (Li)	10
Manganese (Mn)	10
Molybdenum	10
Nickel	10
Phosphorus	90
Platinum	10
Rubidium	10
Selenium (Se)	10
Silicon	10
Silver (Ag)	10
Strontium (Sr)	10
Sulphur (S)	44,700
Tantalum	10
Thorium	10
Tin	10
Titanium	10
Tungsten	10
Yttrium	10
Zinc (Zn)	10
Zirconium	10

Mount Olympus Water Analysis

	Parts per Million
Calcium (Ca)	8
Magnesium (Mg)	4.8
Sodium (Na)	3
Potassium (K)	.70
Chloride (Cl)	4
Chromium (Cr)	.01
Copper (Cu)	.002
Fluoride (F)	.29
Iron (Fe)	.078
Lead (Pb)	<.001
Manganese (Mn)	<.001
Mercury (Hg)	<.0002
Selenium (Se)	<.001
Silver (Ag)	<.001
Zinc (Zn)	.015
Carbonates (CO_3)	<.01
Sulphates (SO_4)	23
Nitrates (NO_3)	.15
Phosphates (PO_4)	.020
Cyanide (CN)	<.01
Ammonium (NH_4)	<.001
Silica (SiO_2)	.5
Phenols	.001
Total Hardness, as $CaCO_3$	40

Rocky Minerals Brand Water Water Analysis

	Parts per Million
Calcium (Ca)	1500.9
Magnesium (Mg)	280.8
Sodium & Potassium (Na & K)	112.4
Chloride (Cl)	32
Copper (Cu)	.42
Fluoride (F)	4.3
Iron (Fe)	1467.3
Manganese (Mn)	173
Silica (Si)	183
Zinc (Zn)	3.37
Sulphate (SO_4)	7360
Total Dissolved Solids	12,120
pH (Hydrogen Ion Concentration)	2.8

Mount Olympus Waters, Inc.

(Salt Lake City)

Located in Neff's Canyon about 5,500 feet up Mount Olympus, in Salt Lake County, is a clean mountain spring. Little change in the quality of water has been found in 50 years of chemical analysis.

Salt Lake City's first bottled water business was established in 1899 and was known as the Mount Olympus Fruit and Livestock Farm Company. The name was subsequently changed to the Mount Olympus Spring Water Company. Then in 1969 Mount Olympus launched its operation in the Northwest; with this expansion the company's name was modified to Mount Olympus Waters, Inc.

Rocky Minerals Brand Water

(Big Rock Candy Mountain)

"And the lemonade springs
Where the blue bird sings,
In the Big Rock Candy Mountain."
— "Haywire Mac" Harry McClintock

The name given the springs by sixteen-year-old "Haywire Mac" referred to in his vintage 1897 ballad (later made famous by Burl Ives) referred to the tang and flavor of the spring's highly concentrated mineral water when mixed with regular drinking water. Today its praises are being sung as a natural tonic containing "all the minerals declared essential for human nutrition, balanced by nature and pure — nothing added, nothing taken away."

Water from the Lemonade Springs has been bottled for 50 years, and used to heal and energize, archeological discoveries indicate, before Columbus — even before the birth of Christ.

In 1974, after three years of negotiation, John H. Gledhill purchased Big Rock Candy Mountain with its Lemonade Springs. By 1984, his newly-founded company, Rocky Minerals, Inc., began a vigorous marketing campaign which has made the Lemonade Springs water available nationwide and worldwide and backs up Rocky Mineral Brand Water with a "Full Satisfaction Guarantee."

VIRGINIA

Seawright Springs

(Mount Sydney)

Early explorers of Virginia learned from Indians of a spring in a certain secluded valley among the hills, which was considered "great medicine for the sick." The Indians called the spring a name which meant "good health water." They claimed the water was different from any other in the region because the Great Spirit had put "great medicine" into it.

This was Seawright Springs located at Mount Sydney in Augusta County, in the quiet hills of the Shenandoah Valley. In 1741, 34 years before the Declaration of Independence, King George II of England deeded by land grant 800 acres of the Shenandoah Valley to a John Seawright.

George Seawright, his last descendant, died in 1888. Eighty-nine acres of his estate, including the spring, were sold for $2,625. By 1890 the spring water had become well known as a healing and restorative agent. The property was then sold to a new owner who expanded the use of bottled spring water by presenting its merits to physicians in various cities. When the property was sold again in 1898, construction was started on a large resort hotel. It was

Seawright Springs Water Analysis

	Parts per Million
Calcium (Ca)	62.4
Magnesium (Mg)	28.2
Sodium (Na)	.3
Chloride (Cl)	5
Copper (Cu)	0
Fluoride (F)	.23
Iron (Fe)	.05
Manganese (Mn)	.02
Mercury (Hg)	.0007
Bicarbonates (HCO_3)	319.6
Carbonates (CO_3)	0
Sulphates (SO_4)	1.9
Nitrates (NO_3)	6.7
Phosphates (PO_4)	0
Silica (SiO_2)	11.3

completed in 1905 and was very popular for two years before it was completely destroyed by fire.

The property changed hands several times during the ensuing years, with the spring water gaining widespread recognition. Accounts receivable, which still exist from the 1920s, show shipments at one dollar per gallon f.o.b. The spring shipped to customers in Fort Worth, Boston, New York City, Cleveland, Chicago, Los Angeles, Tucson and many other cities.

In 1965 Seawright Springs was purchased by I.H. Baker of Stanton, Virginia and S.D. Baker of Arlington, Virginia, who own it today.

The original source of Seawright Spring Water is high in the Blue Ridge Mountains of Virginia. It is a mild mineral water which flows from a depth of over 1000 feet, at a rate of more than one million gallons per day. The flow is unaffected by drought or rain and remains all year at a constant 59° F. Its excellent mineral and chemical content never varies. The water is delivered to distributors in 56,000 gallon tank trucks in both Maryland and Virginia.

WEST VIRGINIA

Tyler Mountain Water Company

((Charleston)

The Tyler Water Company was started in 1930 and has been in continuous operation at the same location ever since. The source is a drilled well, 108 feet deep. Tyler water is sold in six surrounding states, packaged in gallon and eight ounce plastic bottles. The eight

Tyler Mountain Water Analysis

	Parts per Million
Calcium (Ca)	10
Magnesium (Mg)	2.19
Sodium (Na)	2.4
Potassium (K)	.81
Arsenic (As)	.005
Chloride (Cl)	4
Copper (Cu)	0
Fluoride (F)	.78
Iron (Fe)	0
Manganese (Mn)	0
Zinc (Zn)	0
Sulphates (SO_4)	6.5
Nitrates (NO_3)	.6
Surfactants (Detergent)	87
Alkalinity (M.O.), as $CaCO_3$	22
Total Hardness, as $CaCO_3$	34
Turbidity	.05
pH (Hydrogen Ion Concentration)	7.4

ounce bottle is unique in that it is packaged and sold primarily to be consumed by coal miners while working in the mines.

WISCONSIN

Chippewa Springs Corporation

(Chippewa Falls)

Chippewa Springs were discovered by the Ojibwa Indians, a large tribe of the Algonquin family, divided in the Lake Superior region between the United States and Canada.

The water is bottled at the source and has a low sodium content. Two kinds of water are produced; one is pure spring water, the other is artificially carbonated. Chippewa Springs water has a low .005 percent of dissolved minerals.

Chippawa Springs Water Analysis

	Parts per Million
Calcium (Ca)	20
Magnesium (Mg)	12
Iron (Fe)	.03
Manganese (Mn)	.03
Sulphates (SO_4)	3
Silica (SiO_2)	17
Total Alkalinity, as $CaCO_3$	5
Total Hardness, as $CaCO_3$	32

WYOMING

Rock Crest Water Company

(Cheyenne)

Rock Crest Water Company was established in 1933 in Cheyenne. In 1958 an expansion program began, and the water is now bottled in five gallon bottles and cases of six half gallon jugs. Water is drawn from a deep artesian well. Rock Crest uses a demineralization process to remove all minerals from the water, which then contains no chemicals, chlorine, copper sulfate, or aluminum. A good balance of fluorine and selected other minerals are re-injected into the water.

Rock Crest Water Analysis

	Parts per Million
Calcium (Ca)	34
Magnesium (Mg)	5
Sodium (Na)	10
Chloride (Cl)	2
Fluoride (F)	1.6
Bicarbonates (HCO_3)	140
Sulphates (SO_4)	4

International Bottled Waters

CANADA

Naya Spring Water

(Mirabel, Quebec)

Drawn from an aquifer 250 feet deep that is protected by 20 million square feet of wilderness area surrounded by a National Park, Naya Spring Water is naturally bacteria free and therefore can be processed directly from source to bottle. The source is a Cambrian aquifer, created 200 years ago; its waters filtered through layers of limestone, shale and sand, deep into the Laurentian Mountains in what is now Mirabel, Quebec, Canada. The salt-free still water contains a low level of dissolved solids — 200 ppm — and has done well in taste tests. Protected by stainless-steel piping from the source to the plant, Naya Spring Water is packaged automatically without making contact with ambient air into PET (polyethylene terephthalate) bottles, which were molded on the site and filled at the rate of eight bottles per second.

The company is young — taking its name from mythical gods who were the legendary protectors of pure spring water — and business-minded enough to work out an impressive distribution network that thoroughly covers the eastern part of the U.S. and is stretching westward to Arizona.

Naya Spring Water
Water Analysis

	Parts per Million
Calcium (Ca)	39.2
Magnesium (Mg)	22
Sodium (Na)	6.5
Potassium (K)	2.5
Fluoride (F)	.17
Zinc (Z)	.07
Sulphates (SO_4)	10
Total Hardness as $CaCO_3$	199
Total Dissolved Solids	203
pH (Hydrogen Ion Concentration)	7.8

Under the direction of Dr. Malek Basbous, a specialist in water science, the company is developing a technologically advanced bottling plant in Southern California. From this and other facilities planned for the United States, Naya's U.S. Company, Natra Tech Inc., will offer private label bottling of high quality water at lower cost.

Sparkling Spring Water Limited

(Valley, Nova Scotia)

Sparkling Spring Water was established in 1971 at Valley, about six miles northeast of Truro, Nova Scotia. The source was found by researching over 200 known sources of good drinking water in cooperation with officials of the water authority and the Department of Mines in Nova Scotia. The artesian source found at Valley runs continually, at 41° F. The water is bottled as it comes from the earth.

Sparkling Spring Water also produces a demineralized water called "MIN-X", as well as distilled water. The water is distributed throughout Canada's maritime provinces, which include Nova Scotia, New Brunswick and the Prince Edward Islands.

Sparkling Spring Water Analysis

	Parts per Million
Calcium (Ca)	14.4
Magnesium (Mg)	10.1
Sodium (Na)	14
Potassium (K)	.85
Boron (B)	.10
Chloride (Cl)	21
Fluoride (F)	.25
Iron (Fe)	trace
Lead (Pb)	<10
Manganese (Mn)	<.001
Zinc (Zn)	1.5
Sulphates (SO_4)	7
Nitrates (NO_3)	1.05
Silica (SiO_2)	14.4
pH (Hydrogen Ion Concentration)	7.9

St. Justin Mineral Water

(St. Justin, Quebec)

A sparkling mineral spring surfaces in the town of St. Justin, Quebec, which takes the same name as the town. The company is one of the oldest in Quebec, having been in commercial operation since 1895. The water supply is located about 70 feet underground and is brought to the surface through a force pump next to the bottling plant.

St. Justin Mineral Water Analysis

	Parts per Million
Calcium (Ca)	3
Magnesium (Mg)	5.8
Sodium (Na)	500
Potassium (K)	6.2
Arsenic (As)	0
Chloride (Cl)	250
Copper (Cu)	.15
Fluoride (F)	4.2
Lead (Pb)	0
Zinc (Zn)	.05
Bicarbonates (HCO_3)	1650
Sulphates (SO_4)	0
Nitrates (NO_3)	0

Pure, clear water — gift of the earth.

MEXICO

Manantiales Peñafiel
(Tehuacan)

The waters of Tehuacan have been famous since the fifteenth century. They fall from the Citlaltepetl Mountains, 18,100 feet above sea level. Aztec emperor Montezuma took the water for its curative powers, and to give him strength to battle Cortés. Today the mineral water is still used for hydrotherapy, and Mexican researchers claim it soothes arthritis, cleanses the kidneys and removes kidney and gall stones.

The bottling company was established in 1928 by the former Mexican President, Abelardo Rodriguez. The bottling plant is powered by its own hydroelectricity.

Manantiales Peñafiel Water Analysis

	Parts per Million
Calcium (Ca)	574
Magnesium (Mg)	198
Sodium (Na)	333
Potassium (K)	19
Chloride (Cl)	299
Bicarbonates (HCO_3)	1306
Nitrates (NO_3)	10.8
Phosphates (PO_4)	.08
Silica (SiO_2)	52.80

ANDORRA

Aigua d'Andorra

High in the Pyrenees Mountains in the Principality of Andorra is one of the oldest mineral water springs in Europe. The spring

Aigua d'Andorra Water Analysis	Parts per Million
Calcium (Ca)	25.6
Magnesium (Mg)	2.7
Sodium (Na)	1.0
Potassium (K)	.3
Aluminum (Al)	0
Chloride (Cl)	.7
Iron (Fe)	<.1
Manganese (Mn)	0
Bicarbonates (HCO_3)	75.7
Carbonates (CO_3)	0
Sulphates (SO_4)	12
Nitrites (NO_2)	0
Nitrates (NO_3)	1.9
Phosphates (PO_4)	0
Ammonium (NH_4)	<.1

bubbles out through a large rock base surrounded by other smooth rocks in a semicircle, at an altitude of approximately 5,200 feet. One of its outstanding characteristics is that it flows at the same volume and temperature throughout the year. The output is approximately 2.5 million liters per day.

Societat Andorrana d'Aiguas Minerals, S.A., established a bottling plant at the spring in 1973, and installed protective fencing. The water is sold within the region and in other countries in 1.5 liter, .75 liter, and in .33 liter bottles.

Belgium's Spa Reine Spring Water is available in the PVC bottles so familiar in Europe.

BELGIUM

Spa Reine

(Fagnes)

In Fagnes, natural springs surface at an altitude of 1320 feet, producing a water very low in minerals and bottled as Spa Reine. Because of its low salt content, the company claims it is helpful in eliminating wastes and purifying the system.

Spa Reine has one of the most advanced bottling plants in Europe. Covering almost 14 acres, the plant is largely automatic and has a daily production of almost two million bottles.

Spa Reine Water Analysis	Parts per Million
Calcium (Ca)	3.33
Magnesium (Mg)	1.8
Sodium (Na)	2.5
Potassium (K)	.5
Aluminum (Al)	.1
Chloride (Cl)	2.7
Iron (Fe)	.1
Lithium (Li)	trace
Manganese (Mn)	.1
Bicarbonates (HCO_3)	10.65
Sulphates (SO_4)	5
Silica (SiO_2)	7
pH (Hydrogen Ion Concentration)	5.6 to 6

ENGLAND

Schweppes Malvern Water
(London)

Schweppes Limited began bottling water from the Primeswell Spring, also known as the Pewtress Spring, in 1896. The company had bottled water from the Holy Well Spring from about 1850. Schweppes Malvern water was first used medicinally by Dr. John Wall. He stated, "Malvern Water is famous for containing nothing at all."

The source is located on the western slope of the Malvern Hills at an altitude of 700 feet. The water is meteoric and rises through a nearly horizontal fault bound on the bottom by limestone and on the top by Precambrian rock. The Malvern Hills stretch for nine miles which includes the town Great Malvern where a holiday spa is located.

Scheppes Malvern Water Analysis

	Parts per Million
Calcium (Ca)	40
Magnesium (Mg)	14
Sodium (Na)	12
Potassium (K)	1
Chloride (Cl)	35
Carbonates (CO_3)	60
Sulphates (SO_4)	30
Nitrates (NO_3)	5
Silica (SiO_2)	10

FINLAND

Oy Hartwall Ab

(Helsinki)

Oy Hartwall Ab
Raw Water Used in Preparing
Vichy, VIP, and
English Soda Water
Analysis

	Parts per Million
Aluminum (Al)	.04
Chloride (Cl)	50
Fluoride (F)	.3
Iron (Fe)	0
Manganese (Mn)	0
Nitrites (NO_2)	.02
Nitrates (NO_3)	4.7
Total Hardness, as $CaCO_3$	5.8

Oy Hartwall Ab was founded in 1836. This family owned enterprise has been producing artificial mineral waters ever since. At first they were imitating health waters, like those found in continental Europe. Later on they developed waters lower in mineral salts—more like the table waters of the United States.

Wells are drilled up to 120 meters deep, and the well water is blended with municipal water. It is then treated with ferrosulfate, calcium hydroxide, and sodium hypochlorite, filtered through sand and active carbon, and bottled with ozone gas. Three different kinds of water are sold. "Vichy" water is rich in bicarbonates, "VIP" is a still water low in mineral content, and "English Soda Water" is a sparkling water with less salts than Vichy.

Pyynikki
(Tampere)

In southwestern Finland the Pyynikki Company produces three mineral waters. The water source is located in a glacial gravel hill which offers efficient filterage and a natural mineral balance. The original wells were dug back in the 1890s. Beginning in 1950, new wells were dug at 94 meters, 114 meters and 180 meters. The three kinds bottled are: "Vichy," a very effervescent mineral water with large quantities of inorganic salts; "Soda Water 114," a slightly alkaline and sparkling water used predominantly as a mixer; and "Kivennaisvesi," an alkaline and lightly sparkling table water.

The old drinking hall at the Evian spa.

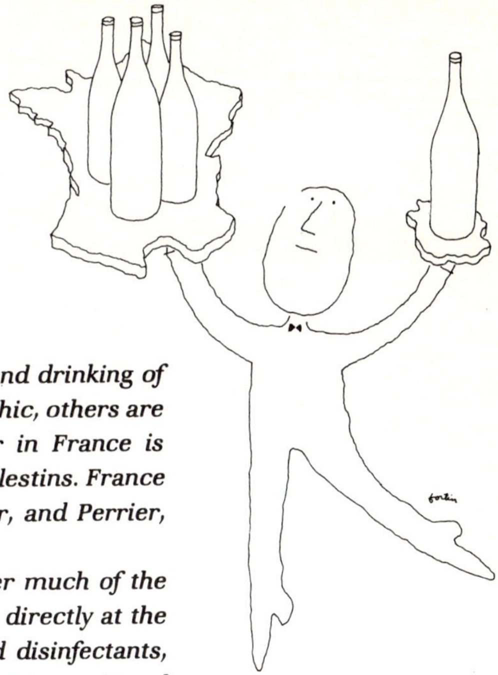

French Mineral Waters

As with so many other things in France, the bottling and drinking of mineral water has become an art. Certain waters are chic, others are rare, but all are unique. The appreciation of water in France is centuries old, dating back to Julius Caesar at Vichy Célestins. France is also the home of Evian, the world's largest bottler, and Perrier, America's most popular imported water.

The French government holds strict control over much of the bottling process. For example, water must be bottled directly at the source. Bottles cannot exceed the two liter size and disinfectants, such as chlorine and ozone gas, may not be added. Most mineral water carries the government seal "d'interêt public," which means its healthful properties are recognized by the state. The government keeps a watchful eye on water properties and in many cases analyses are done daily.

Throughout France water is available in glass as well as in plastic bottles, the latter being far more popular (plastic is disposable while glass bottles are sold only with a deposit; also, plastic is cheaper, lighter to transport and less likely to break). However, glass bottles are still used, especially for exportation to the United States, as U.S. law prohibits the importation of plastic bottles. French scientists have experimented with various plastics and have found that certain mineral waters act differently on plastics. Through trial and error they have found which plastics are compatible with various mineral waters. Some waters, such as Perrier, cannot be bottled in plastic because of the high pressure of the gas. Evian was the first bottler to market water in aluminum cans, an operation they have since ceased due to the high cost.

FRANCE

Badoit
Water Analysis

	Parts per Million
Calcium (Ca)	355
Magnesium (Mg)	198
Sodium (Na)	205
Potassium (K)	16
Arsenic (As)	.01
Chloride (Cl)	64
Copper (Cu)	trace
Fluoride (F)	1.2
Iron (Fe)	.1
Lead (Pb)	trace
Lithium (Li)	1
Manganese (Mn)	.9
Sulphates (SO_4)	43.5
Nitrites (NO_2)	8.0
Nitrates (NO_3)	trace
Silica (SiO_2)	34.5
Free CO_2 gas	2000
Alkalinity in cubic centimeters of acid N/10	290
Alkalinity, as CaO	812
Organic Matter in Oxygen	.35

Badoit

(St. Galmier)

In the valley of the Loire, at St. Galmier, the naturally sparkling Badoit water comes bubbling up. St. Galmier dates back to the thirteenth century. In 1539 the water was described by Gilles Gorrozet as "perhaps more delicious than beautiful wine." Naturally sparkling, with a moderate fluoride content, the water comes from a deep rock spring on the edge of the Monte du Lyonnais which is protected by the French government. One of the rare features at Badoit is the fact that the entire bottling room is maintained at the same pressure as the water in the spring. In this way, Badoit retains all its natural effervescence when bottled.

A letter requesting a shipment of Badoit water, written by Louis Pasteur in 1887.

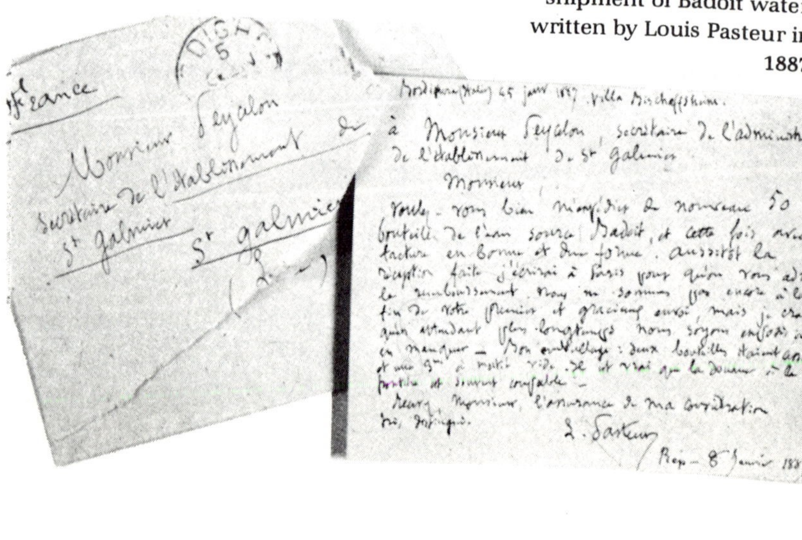

The Vierge du Pilier and l'Hôtel de Ville at St. Galmier. St. Galmier is the home of Badoit.

Contrexéville

(Vosges)

On the border between France and Germany in a range called the Faucille Mountains is the Contrexéville Source Pavillon Resort. Here, 1,148 feet above sea level, pioneering climbers discovered the source around 1760. The Pavillon Spring brings to the surface close to 44,000 gallons of water a day.

The original spa was built over a hundred years ago in Greco-Roman style. The springs were taken over by the Germans during World War II, and most of the spa and thermal buildings had to be rebuilt.

The spa is still a sociable resort, similar to those of centuries ago. Rest cures are very popular, although visitors find the waters invigorating enough to allow them to engage in a few hands of baccarat at the Casino in the center of the park. The water is known for its powerful diuretic nature because of a high calcium sulphate content.

Contrexéville Water Analysis

	Parts per Million
Calcium (Ca)	546
Magnesium (Mg)	44.5
Sodium (Na)	9
Potassium (K)	3
Chloride (Cl)	10
Iron (Fe)	.2
Bicarbonates (HCO_3)	366
Sulphates (SO_4)	1190
Nitrites (NO_2)	0
Nitrates (NO_3)	0
Phosphates (PO_4)	.15
Ammonium (NH_4)	0
Silica (SiO_2)	8.5
Dry residue at 105/110° C.	2,290
pH (Hydrogen Ion Concentration)	7

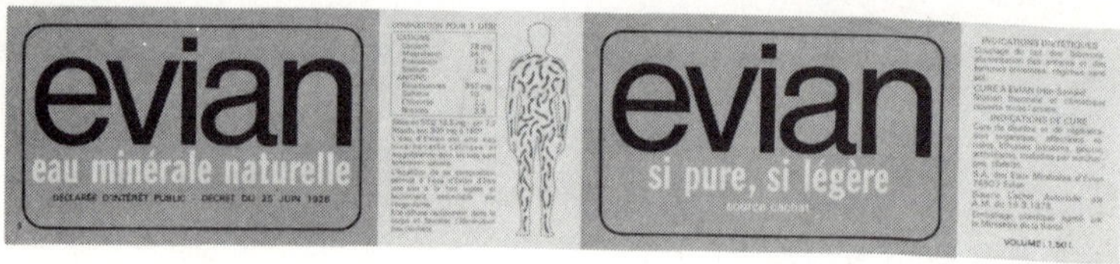

Evian

At the foot of Mont Blanc lies the lakeside health resort of Evian. Melted snow and ice from the mountain are filtered down through the Alps and glacial rocks, and become enriched with mineral salts. The spa itself was famous almost 200 years ago. The town basked in the glory of the spa's popularity in the early 1900s when Lake Geneva was the place to see and be seen. Casinos were built around the lake helping visitors lighten their pocketbooks.

The French Ministry of Health declared Evian water to be *d'interêt public.* The Pasteur Institute has continued to make regular checks of the water and its composition. In fact, the water has scarcely varied in physical-chemical composition since the first analysis over a hundred years ago.

In 1987, Evian bottled a little over 900 million liters, with their bottling line running at 25,000 bottles per hour.

They recently introduced a two liter, square bottle for the French market. Evian is also involved in the marketing of other beverages, including an electrolyte replacement beverage called Athlon which is currently being distributed in Europe.

"Water of high quality does not need sterilization, as with ozone, which is commonly used in the United States," said Vincent Ducasse, Evian's director of research and development. "If water is consistent, then it is *A Water*; if it is not consistent then it is water."

Evian
Water Analysis

	Parts per Million
Calcium (Ca)	78
Magnesium (Mg)	23
Sodium (Na)	5.5
Potassium (K)	.75
Chloride (Cl)	2.2
Bicarbonates (HCO_3)	357
Sulphates (SO_4)	10
Nitrates (NO_3)	3.8
pH (Hydrogen Ion Concentration)	7.18

Manually capping and labeling the bottles in the original Evian plant, c. 1900.

A luxury at the spa — showering in Evian water. The Evian water cure begins with two glasses in bed at sunrise, followed by two more glasses at noon and one preceding dinner. The cure brings about increased urine discharge, thus removing toxins from the body.

Perrier
(Vergèze)

Man's acquaintance with the sparkling waters of Perrier predates the emergence of France by centuries. The spring was believed to have been the resting place of Hannibal in 218 B.C. In the first century A.D. it was the site of Roman baths, enjoyed by the legions which built the aqueducts and the Temple of Diana—which still stand 15 kilometers away in Nîmes.

The bottling company's beginning makes an interesting story. One sunny day at the turn of the century, a young Englishman hiked across the stony plains of Nîmes with his tutor. His name was A.W. St. John Harmsworth (donor of the Harmsworth Trophy in racing). His brother, publisher of the London *Daily Mail,* had sent him to France to study the language.

Hiking with him and the tutor was an elderly gentleman with a small white beard, a Dr. Perrier. Perrier was a collector. While others might collect butterflies, Perrier collected mineral springs, and he had just made an important addition to his collection, a spring eight miles from Nîmes, between the extinct volcanoes of Auvergne and Vergèze. This was the same spring that another traveler, two centuries earlier, had gazed upon in awe, writing, "Imagine a spring that bubbles just as if it were boiling while its waters kept cold." In 1863, the Emperor Napoleon III first decreed that the naturally sparkling waters of the spring, then owned by Dr. Perrier, should be bottled "for the good of France." And it was to this spring (then known as Les Bouillens) that Perrier led Harmsworth and his tutor.

By the time Harmsworth reached the spring his throat was parched from the dusty road and he knelt down and drank deeply from sparkling icy waters. He was so delighted with their refreshing quality that he turned spontaneously to Perrier and said, "Doctor, I will buy your spring. I promise you it shall always bear your name,

Perrier
Water Analysis

	Parts per Million
Calcium (Ca)	140.2
Magnesium (Mg)	3.5
Sodium (Na)	14.0
Potassium (K)	.6
Chloride (Cl)	30.9
Bicarbonates (HCO_3)	347.7
Sulphates (SO_4)	51.4
Nitrates (NO_3)	0

and we will bottle the waters and sell them around the world!"

Eleven years after buying the springs Harmsworth was crippled in an automobile accident on the road to Toulouse. However, this did not thwart his interest in the Perrier Spring, and he guided its operations from a wheelchair. Being a former football player, tennis champion and gymnast, he decided to give the Perrier bottle its distinctive shape, that of an Indian club. He also coined the now famous slogan, *Perrier, le champagne des eaux de table* (Perrier, the champagne of table waters). Harmsworth's dedication to Perrier was boundless. Near the spring he built an English style house for himself, on land bought from André Gide's mother, and created a park boasting rare and exotic plants. He came to this home near the Perrier factory many times a year.

Harmsworth's faith in the Perrier spring was justified, for at the time of his death in 1933, Perrier production exceeded ten million bottles a year. Some feared Harmsworth's boldness concerning Perrier water and the future. However, time proved them wrong, for 15 years after Harmsworth's death, changing habits around the world in eating and drinking created a five-fold increase in demand for Perrier water.

Perrier and lime has been nicknamed the "power drink." It is popular with many athletes after strenuous exercise, and has been the official drink of the Tour de France bicycle race for 30 years. Sparkling water is absorbed into the tissues more rapidly than still water.

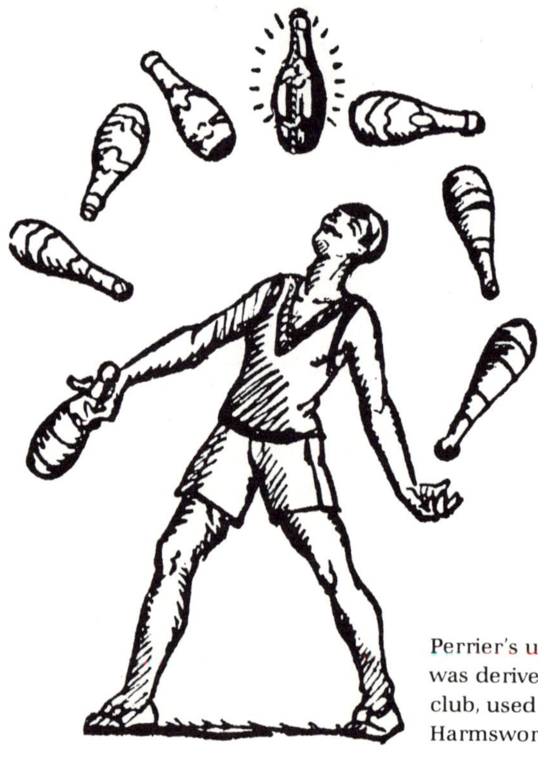

Perrier's unique bottle shape was derived from the Indian club, used by founder Harmsworth.

A.W. St. John Harmsworth. From the oil portrait which still presides over the conference room at Perrier.

By 1949, Harmsworth's enterprise had become a French company known as the "Société de la Source Perrier." Yet Harmsworth's ideas remained its guiding spirit, and because of this, Perrier kept production apace with demand.

Perrier is a most unusual mineral water, because it has become immensely popular without the addition of a famous spa. More than 360 million bottles are now consumed each year.

Source: Deep below the plains and vineyards of southern France, in a process which began over 100 million years ago, gases trapped in volcanic eruptions of the Cretaceous era are naturally released. They travel up toward the surface through cracks and fissures of porous limestone and the cracked marls above it. As they rise they meet and mingle with icy, crystalline mineral waters of exceptional purity and clarity. The phenomenon of Perrier's natural carbonation takes place underground. The waters then break a single surface, the source of Perrier. The source is kept pure by the natural, nonporous clay barrier beneath the spring, which stretches for several kilometers in all directions and prevents surface waters from seeping down to contaminate the bubbling spring. A deep layer of white sand topped with sand and gravel acts as a basin in which the carbonated water collects beneath the spring, and also acts as a natural filtration system through which the water rises. The French government has made it a law that no wells of any kind may be dug within five kilometers of the source.

Vichy Célestins

(Vichy)

Vichy lies in the heart of France. Vichy Célestins has an unusual salty flavor, unique as is the history behind the spring. It is believed that in prehistoric times the Célestins Spring one day burst through the earth's surface with considerable force, creating a spectacular geyser. As the water fell to earth and flowed over the ground in all directions, it left deposits which formed horizontal rock layers of calcium carbonate. The rocks, which took thousands of years to form, were eventually eroded, and today only fragments of these rocks still stand.

Vichy's strata of aragonite rock was formed by water borne mineral deposits over a period of millions of years.

Vichy Célestins Water Analysis

	Parts per Million
Calcium (Ca)	96
Magnesium (Mg)	10.3
Sodium (Na)	1.33
Potassium (K)	75
Chloride (Cl)	254.4
Iron (Fe)	.1
Bicarbonates (HCO_3)	3385.5
Sulphates (SO_4)	141.2
pH (Hydrogen Ion Concentration)	6.61

Célestins is not the only spring to bubble out of Vichy. Four other natural springs include Puits Carré (or Chamel Springs), the hottest, which surfaces at 42.5° C. It is used at the Palais des Sources for baths, along with water from the Luccas Spring and Grande Grille. L'Hôpital Spring is popular at the drinking fountains of the spa. The only other known bottled water from the region is Vichy St. Yorre which is from the man-made St. Yorre Spring, outside the government protected zone.

Julius Caesar built a health resort at the site of Vichy 2,000 years ago, complete with temples and bathing pavilions. Destroyed by barbarian invasions and rebuilt several times, the spa was last rebuilt by Napoleon II. Napolean III drank the waters and endowed the town with many improvements. In 1864 the State built the

Casino Teat and the famous Restauration (the largest open-air café in the world).

Suddenly, one day after the turn of the century, the spring dried up. Holes were bored, and the Célestins Spring resurfaced, more powerful than ever due to a force pump.

The calcium carbonate or aragonite rock from which the water gushes stands in a plate glass frame. It overflows onto marble shelves and finally into stone cups where the water is served in glasses. The bottling plant which produces Vichy Célestins stands in an old grotto. More than eighty million bottles of Vichy Célestins are sold every year.

Contrary to popular belief, Vichy water is not a laxative—in fact it helps to prevent diarrhea.

The Grand Etablissement at Vichy Célestins.

Vittel Grande Source

(Vittel)

Vittel Grande Source
Water Analysis

	Parts per Million
Calcium (Ca)	202
Magnesium (Mg)	36
Sodium (Na)	3
Potassium (K)	2
Fluoride (F)	.4
Bicarbonates (HCO_3)	402
Sulphates (SO_4)	306
Nitrates (NO_3)	5
Nitrites (NO_2)	0
Total Dissolved Solids	1014
pH (Hydrogen Ion Concentration)	7.1

The Romans were first to discover the beneficial properties of the water gushing from Vittel springs. Fragments of ancient statues from their baths, and the temple they built in honor of the god Grannus, are still being recovered on grounds of the world-class spa and resort that Vittel has developed around the springs, and operates in cooperation with Club Med. In 1853, Louis Bouloumie, a physician and lawyer who recovered from an "incurable" health condition after drinking its waters, purchased the springs and began building the spa that quickly attracted the European elite.

Vittel Grande Source was officially declared to be *d'interet public* on December 29, 1903. The French Minister of Public Health recognized Vittel's health-giving properties when they were accredited by the National Medical Academy. Tests in 1979 indicated that Vittel water had both preventive and therapeutic qualities in controlling kidney disease and gout. Vittel continues to promote the health-benefiting aspects of its magnesium- and calcium-rich waters, publicizing recent research highlighting the importance of these minerals in supporting overall health.

Besides Grande Source, Vittel bottles Abatilles (a mineral water from the south of France), Pierval (a spring water from Normandy), and flavored water called Vittelloise. They also bottle Diabolis (a line of soft drinks), Ricqles (a mint drink), and Delices (a fruit drink).

Vittel bottles in glass and plastic bottles under the supervision of the European Communities Commission which requires that

"The Source." Vittel spring is sheltered by this enclosure.

Vittel toasts its U.S. success in California.

Vittel, an island of healthful elegance in France.

L'Ermitage and golf course at the Vittel spa.

no plastic (vinyl chloride monomer) migrates from the container into its water. Vittel sells in excess of one billion bottles of mineral water a year in over 70 countries, and is the world's largest bottler of still mineral water. Seventy percent of Vittel is bottled in plastics, 17 percent in glass and 13 percent in other containers, such as cups.

Vittel Grande Source bottling plant has five lines which bottle a total of 22,000 bottles per hour. Since the beginning of 1987, Vittel introduced robots, made by Jeumont-Schneider, called Servocar. These robots move cases of water from the bottling line to storage and from storage to the loading docks. They follow an invisible electric-magnetic strip embedded in the concrete floor, and greatly reduce labor costs.

Asked what constitutes a good water, Vittel's medical director, Dr. Anne-Marie Delabroise, replied: "A good water is protected from pollution; so it is free of contaminants. It should be chemically and biologically clean and not need filtration. A good water does not alter chemically throughout the year; it has constant temperature and flow. A good quality water has no sodium or fluoride. A natural mineral water should not be pumped. If it is pumped, it pulls in the drainage from the surrounding area, and the security of the underground aquifer is threatened. You can't *make* mineral water; the minerals will precipitate because mineralization takes time. In the case of Vittel it takes 25 years from the aquifer to the spring."

In 1985, in honor of French mineral water's debut in this country 200 years ago when Benjamin Franklin imported it at the conclusion of his nine-year tenure as Ambassador to France, Vittel announced that with its purchase of Bartlett Springs it would be offering the American public two distinctive mineral water products and four different flavors — the most extensive selection of mineral waters ever debuted simultaneously by a foreign mineral water company.

Bottled water was first imported to this country by none other than Benjamin Franklin. It seems that Ben was not only an inventor, philosopher, journalist, patriot and diplomat, but also a gourmet. During his term as Ambassador to France from 1776 to 1784, he was introduced to natural spring bottled water. He was so impressed with the spring water that he arranged to have a supply shipped to him before he returned to Philadelphia in 1785. Friends visiting his home enjoyed his famous bottled water.

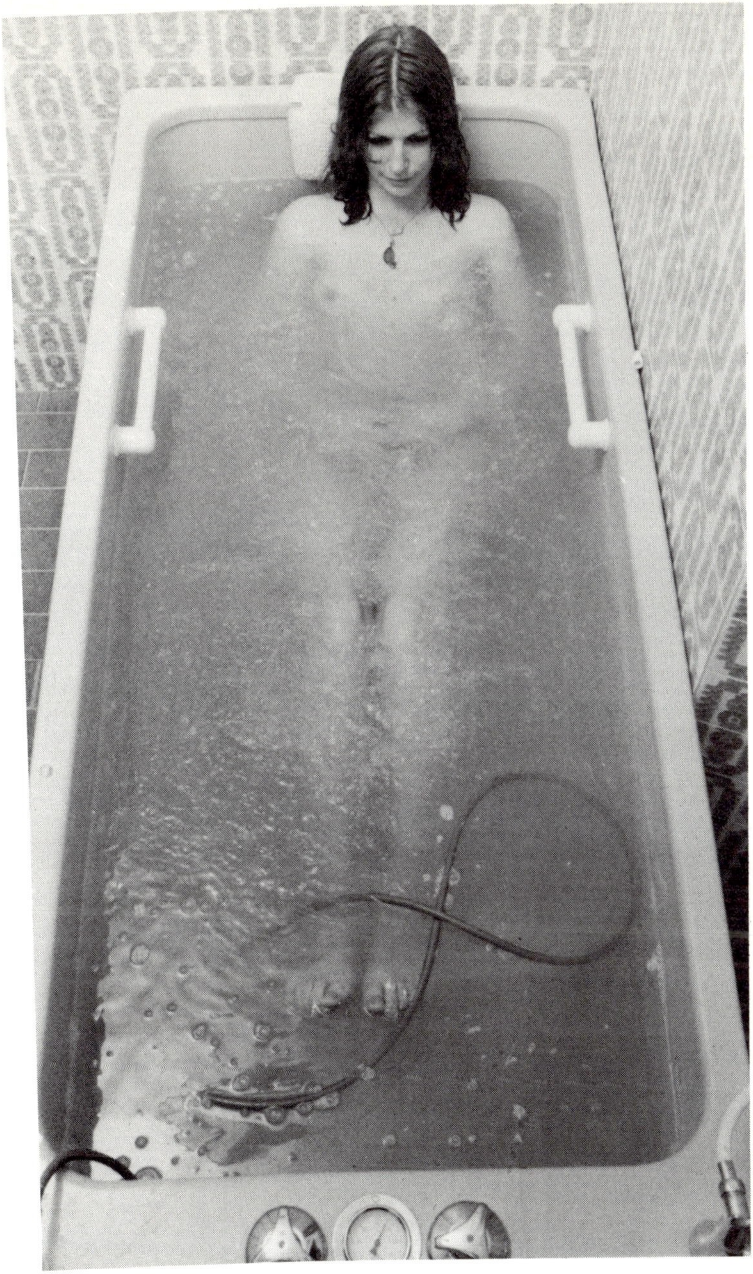

A bubbling bath treatment is part of the water therapy course.

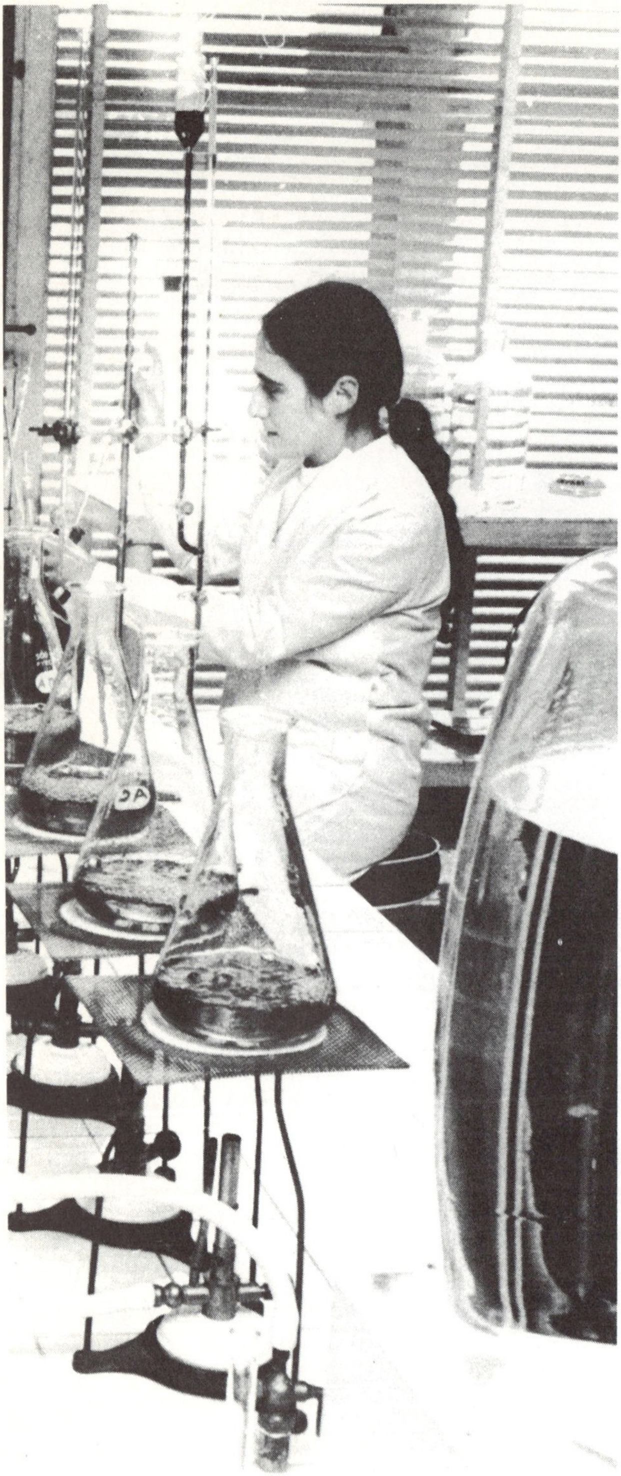

The laboratory at Vittel keeps constant control on water quality.

Volvic

(Puy-de-Dome)

Volvic is a still natural mineral water bottled in the Auvergne, central France. The source is protected by an area of 17 square miles, and, as are all French bottled waters classified as "Eau Minérale Naturelle," the water is bottled at the source. The basin of the Volvic source is an ancient valley filled over many thousands of years with layers of various volcanic rocks.

The water is popular in France and England not only for drinking but also for skin care due to its low mineral content.

Volvic Water Analysis

	Parts per Million
Calcium (Ca)	10.4
Magnesium (Mg)	6
Sodium (Na)	8
Potassium (K)	5.4
Chloride (Cl)	7.5
Bicarbonates (HCO_3)	64
Sulphates (SO_4)	6.7
Nitrates (NO_3)	2
Silica (SiO_2)	40

The Volvic basin, a natural volcanic filter, is an ancient valley filled over many thousands of years with layers of various porous rocks pierced by an infinite number of microscopic holes. From this basin, far below ground, Volvic springs free of impurities due to its filtration through the overlying strata. Volvic is naturally low in minerals and has a perfectly neutral pH (7.0).

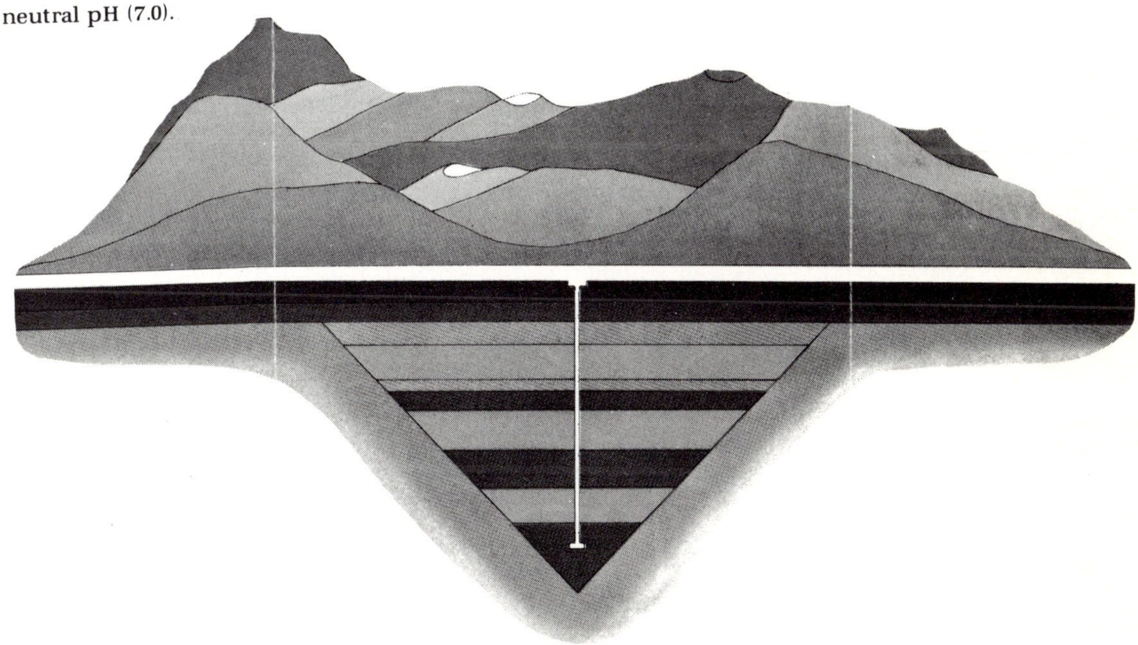

GERMANY

Apollinaris Water Bad Neuenahr

(Bad Neuenahr, West Germany)

The Apollinaris Spring is situated in the beautiful Ahr valley at Bad Neuenahr. The ground surrounding the spring was originally covered with vineyards. In 1852 the vineyard owner noticed that in one particular spot the vines did not grow well. He had chemical experts inspect the land who informed him that the soil was impregnated with carbonic acid and recommended digging a well. At a depth of 40 feet the spring was reached and water gushed to the surface. The name Apollinaris was given to the spring by the owner in honor of a neighboring church dedicated to St. Apollinaris.

Apollinaris Mineral Water was known to benefit catarrhal affections of the throat and respiratory organs. The water when mixed with one-fourth as much boiling milk was said to give general relief. Also in disorders of the urinary tract relief was said to be afforded by the alkaline, solvent and diuretic properties of Apollinaris Water.

The water is highly mineralized with bicarbonates of various elements. In 1970 a new bottling plant was built, capable of producing over a million bottles a day. The water has remained constant since its first analysis. It is one of the world's most popular bottled waters with over 260 million liters bottled in 1977.

Apollinaris
Water Analysis

	Parts per Million
Potassium Chloride	60
Sodium Chloride	358
Sodium Bromide	.2
Sodium Sulphate	255
Sodium Bicarbonate	2,106
Lithium Bicarbonate	5
Calcium Biphosphate	.4
Calcium Biarsenate	.2
Calcium Bicarbonate	466
Magnesium Bicarbonate	857
Ferro Bicarbonate	3
Silicic Acid	26

Staatl. Fachingen

The Romans used the springs for their health-giving properties, and later, the kings, queens, dukes, and clergy of Germany enjoyed its natural mineral waters.

Today, the company offers the water "for the liberation of your genius," alluding to a passage in Goethe's letter to his daughter-in-law Ottilie: ". . . the next four weeks are supposed to work wonders. For this purpose I hope to be favored with

Fachingen water and white wine, the one to liberate the genius, the other to inspire it.''

Currently, as *staatl.* implies, the springs are state-owned, the bottling process carefully supervised.

Staatl. Fachingen Water Analysis

	Parts per Million
Calcium (Ca)	132.1
Magnesium (Mg)	79.8
Sodium (Na)	636
Potassium (K)	17.4
Chloride (Cl)	179
Iron (Fe)	2.3
Manganese (Mn)	.44
Ammonium (NH_4)	.21
Nitrates (NO_3)	.79
Sulphates (SO_4)	50.6
Hydrogen carbonate (HCO_3)	2159
Hydrogen phosphate (HPO_4)	.002
Silicic acid	44.8
Total Dissolved Solids	3302
Free Dissolved Carbon Dioxide (CO_2)	1728
pH (Hydrogen Ion Concentration	61

HUNGARY

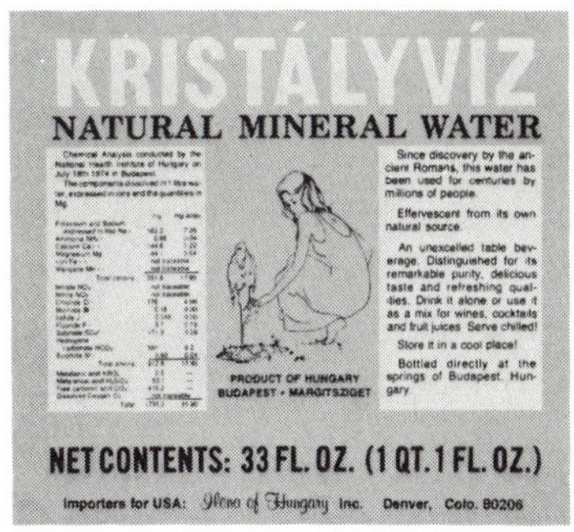

Kristalyviz Natural Mineral Water

(Budapest)

Hungary has been renowned for mineral water springs for centuries. The first baths were built by the Turks over 400 years ago; Hungary has since been a world center for water cures.

In more opulent days the wealthy of Budapest had mineral water piped directly into their homes. Pipes from the springs still exist although the government now restricts the flow of mineral water to two hours a day.

Krystalyviz (crystal water) comes from a source in Budapest which surfaces at 70° C. Empress Elizabeth of Austria used to drink Krystalyviz. In Europe it is still popular to help acid indigestion and prevent kidney stones and is recommended for nursing mothers because of its high calcium content. The Hungarian Ministry of Health recommends the water for consumption by children due to the high fluoride content which builds strong tooth enamel. The water is naturally sparkling.

Krystalyviz Natural Mineral Water Analysis

	Parts per Million
Calcium (Ca)	144.6
Magnesium (Mg)	44.1
Sodium and Potassium	162.2
Bromide (Br)	.18
Chloride (Cl)	176
Fluoride (F)	3.7
Iodide (I)	.048
Iron (Fe)	none
Manganese (Mn)	none
Bicarbonates (HCO_3)	561
Sulphide (S_2)	.6
Sulphates (SO_4)	171.3
Nitrites (NO_2)	none
Nitrates (NO_3)	none
Ammonium (NH_4)	.66
Free CO_2 gas	415.2

ITALY

Acqua Minerale Naturale Sant'Elena

(Chianciano Terme)

Near the ancient ruins of Chapel St. Elena gushes forth the naturally sparkling mineral water called Sant'Elena. Known in Italy for its curative properties, the water was first recorded in the memoirs of Dr. Baldassarre of Siena in 1745.

The source is surrounded by 120 feet of sand and yields an alkaline water. It has been used in Italy to treat urinary tract disorders, and is reported to possess a spasmolytic action on the musculature of the ureter and bladder, a pain relieving action on the nerve endings of the urinary tract, an anti-inflammatory action on mucous membrane of the bladder and kidneys, a cleansing action on bladder catarrh, and general metabolic stimulating properties.

Because of Sant'Elena's special properties, it is recommended in the treatment of urinary calculosis, cystitis and other chronic inflammations of the urinary tract, gout, obesity, arteriosclerosis, edema of liver disease, and excessive water retention.

Chianciano Terme is a popular center for mineral water cures, and the Sant'Elena spring offers visitors a chance to take the waters slightly warm, then stroll in the park or play miniature golf.

Acqua Sant'Elena
Water Analysis

	Parts per Million
Calcium (Ca)	111.72
Magnesum (Mg)	10.19
Sodium (Na)	28.1
Potassium (K)	1.32
Aluminum (Al)	.012
Chloride (Cl)	31.7
Iron (Fe)	.007
Lithium (Li)	.005
Strontium (Sr)	.6
Bicarbonates (HCO_3)	344.6
Sulphates (SO_4)	38.2
Nitrates (NO_3)	19
Silica (SiO_2)	11

Acqua Fiuggi

(Fuiggi)

The source, in the central Italian town of Fiuggi overlooking the Sacco River Valley, was officially discovered in 1696. Michelangelo, Pope Boniface VIII, the Colonna family (of the Battle of Lepanto fame), and many others found solace in the waters of Fiuggi. It is one of the most popular still waters in Italy because of its clear, sweet taste. It can be found in most Italian restaurants in Europe and America.

Acqua Fiuggi
Water Analysis

	Parts per Million
Calcium (Ca)	11.8
Magnesium (Mg)	2.5
Sodium (Na)	6.8
Potassium (K)	.1
Chloride (Cl)	15.2
Copper (Cu)	trace
Iron (Fe)	trace
Lithium (Li)	4
Manganese (Mn)	trace
Strontium (Sr)	trace
Bicarbonates (HCO_3)	32.4
Sulphates (SO_4)	4.5
Nitrates (NO_3)	4.1
Silica (SiO_2)	33
Free Carbonic Anhydride	27
Total Hardness in French degrees	5.4
Total Dissolved Solids	100.8

Drinking Fiuggi water from a fountain at the spa.

San Pellegrino
(Milan)

San Pellegrino has been mentioned in writings as far back as the twelfth century. Its powers were praised by Astori, Baronic, Carrara and others. The water became famous as a curative when Guiseppe Pasta worked out a systemic cleansing method. The bottling company was established in 1899 and a spa was built in the grand Italian manner, lavishly decorated with elaborate painted ceilings and marble floors. It has been popular ever since, and has been visited by many dignitaries.

The water, high in bicarbonates and sulfates, has been used to treat a wide range of physical disorders which include liver diseases, kidney and urinary ailments, and metabolic and digestive disorders. The Pellegrino treatment for home use is as follows:

> Open the bottle the night before so that the fizzy water, the carbonic acid, may evaporate. The morning after the water should be heated in a water bath to the spa temperature which is 80.6° F. During the first week, before rising, sip one pint of water in a one hour time period. Half an hour to one hour later, after rising, take a cup of coffee or warm milk. In the late morning drink another glass of water, then another one in the afternoon, remembering that the water must be lukewarm and drunk in small draughts. During the second week a second glass may be added to the mid-morning dose and a further one during the third week. Three or four weeks should be sufficient for this home treatment, which may well be repeated a few months later.

San Pellegrino may be found throughout the world in most

San Pellegrino
Water Analysis

	Parts per Million
Calcium (Ca)	203.2
Magnesium (Mg)	59.4
Sodium (Na)	44.2
Potassium (K)	4.1
Chloride (Cl)	67.26
Fluoride (F)	.58
Lithium (Li)	.2
Strontium (Sr)	3.55
Bicarbonates (HCO_3)	225.65
Nitrates (NO_3)	.75
Sulphates (SO_4)	560
Free silicic residue (SiO_2)	11.9
pH (Hydrogen Ion Concentration)	7.25

Italian restaurants and specialty stores. It is a delightful and refreshing sparkling water.

A 1921 San Pellegrino poster.

PORTUGAL

Água do Luso

(Luso)

Luso is 600 feet above sea level, and lies at the foot of the Bucaco Mountains in northwest Portugal. The Luso Spa stands above the mineral spring, which produces 25,000 liters an hour. The water is slightly gaseous, crystal clear and perhaps the most popular water in Portugal. High in radioactivity, it is used as a diuretic and for detoxification. The water treatment at Luso is said to cure urinary disorders, arthritis, varicose veins, diabetes and allergies.

Água do Luso
Water Analysis

	Parts per Million
Calcium (Ca)	.76
Magnesium (Mg)	1.42
Sodium (Na)	6.5
Potassium (K)	.91
Aluminum (Al)	.01
Barium (Be)	.01
Beryllium (Be)	.001
Bromide (Br)	.03
Chloride (Cl)	8.95
Chromium (Cr)	.001
Copper (Cu)	.01
Fluoride (F)	.07
Iodide (I)	.001
Iron (Fe)	.005
Lead (Pb)	.01
Manganese (Mn)	.003
Tin (Sn)	.01
Bicarbonates (HCO_3)	4.23
Sulphates (SO_4)	1.88
Nitrates (NO_3)	1.85
Phosphates (PO_4)	.07
Silica (SiO_2)	13
pH (Hydrogen Ion Concentration)	5.5

SCOTLAND

Jaggers Sparkling Scottish Water

(Greenock)

Scots have often accredited their world famous whiskey to their water. Jaggers water has long been used as a mixer with whiskey. Collected from the Peat Moorland Hills at Loch Thom near Greenock, the water is chemically treated through pressure filters with sulfate of alumina and lime, and is chlorinated prior to bottling. Jaggers is sold in one-third liter bottles.

Jaggers Sparkling Scottish Water Analysis

	Parts per Million
Magnesium (Mg)	1.2
Aluminum (Al)	.1
Arsenic (As)	none
Cadmium (Cd)	none
Chloride (Cl)	10
Chromium (Cr)	.001
Copper (Cu)	.17
Fluoride (F)	.05
Iron (Fe)	.06
Lead (Pb)	.03
Manganese (Mn)	.01
Selenium (Se)	.005
Zinc (Zn)	.19
Calcium Carbonate ($CaCO_3$)	14.8
Sulphates (SO_4)	13.1
Thiosulphates (S_2O_3)	none
Nitrates (NO_3)	.40
Cyanide (CN)	none
Silicate (Na_2SiO_3)	.7
Free SO_2 gas	none
Phenols	none
Chloramines	.09
Color	clear

SWEDEN

Ramlösa, Royal Swedish Mineral Water

(Hälsingborg)

Ramlösa is an ancient spa that developed into a meeting place for Scandinavian nobility, and later became a fashionable spa for all of northern Europe. Although the spa itself closed in 1975, the world-renowned Ramlösa drinking water, first bottled in the early 1880s, is currently drawn from spring to bottle at the rate of 100 million bottles a year. Ramlösa holds the distinction of being designated purveyor to His Majesty the King of Sweden and Her Majesty the Queen of Denmark. Carbon dioxide is added to its natural alkaline water.

Ramlösa
Water Analysis

	Parts per Million
Calcium (Ca)	2.5
Magnesium (Mg)	.5
Sodium (Na)	230
Potassium (K)	1.4
Chloride (Cl)	30.0
Copper (Cu)	.02
Fluoride (F)	3.0
Bicarbonates (HCO_3)	537
Sulphates (SO_4)	10.0

Swiss Mineral Waters

Before 1900 the only bottled water used in Switzerland was prescribed by doctors. The Swiss were aware of the popularity of Roman baths and spas that had been used for centuries, and that many of the surrounding countries were producing and drinking mineral waters. Switzerland, however, was not interested in developing her own spring water resources until the 1900s. Some interest was generated before the beginning of World War I, then the war effort became more important and water development was put aside.

By the 1930s so many mineral water sources were discovered they seemed to be mushrooming out of the ground. The year 1932 put bottled mineral water in people's homes —previously it had only been available in restaurants. By 1935 more than 50 different mineral waters were being produced. But it was not until the early 1950s that the Swiss fully realized the beneficial qualities and values of nature's mineral water. Soon there were so many types of natural mineral waters being produced in Switzerland (and the rest of Europe) that artificially carbonated soft drinks lost popularity.

SWITZERLAND

Henniez-Lithinée

(Henniez)

Henniez is the site of ancient Roman baths. One of the Henniez springs, Bonne Fontaine, has been noted on ancient maps of the region. The emperor Vespasien took the waters on his visit to Aventicum. Bonaparte stopped at Henniez to refresh himself while on campaign. At the end of the 19th century the spa was one of the most elegant meeting places in Europe.

In 1880 Dr. Borel, director of the spa, introduced findings which supported the virtues of Henniez water. (The chemist, Ernest Chuard, who worked on the water analysis later became President of the Swiss Confederation.) The bottling plant began operation in 1908.

Aside from Bonne Fontaine, two other sources are bottled, Alcaline and Espérance. The three springs are in the valley of the Broye southeast of the village of Henniez. They produce an alkaline water rich in bicarbonates of calcium and magnesium.

The Swiss use the water to help remedy ailments of the liver, bladder and kidneys. It is a natural diuretic.

The factory, rebuilt in 1968, now produces a half million bottles per day.

Henniez-Lithinee
Water Analysis

	Parts per Million
Calcium (Ca)	98.3
Magnesium (Mg)	18.31
Sodium (Na)	7.49
Potassium (K)	1.07
Aluminum (Al)	.27
Chloride (Cl)	11.5
Fluoride (F)	.06
Iron (Fe)	.08
Lithium (Li)	.13
Strontium (Sr)	1.02
Bicarbonates (HCO_3)	360.13
Sulphates (SO_4)	9.92

Valser St. Petersquelle

(Thermalquellen Vals, Graubünden)

At the thermal resort of Vals in central Switzerland lies the St. Pierre Spring, 1257 meters above sea level. From this spring the renowned Valser St. Petersquelle is bottled. The water, rich in sulphur, calcium and magnesium, is drunk by the Swiss to aid digestion and to stimulate the lower intestinal tracts. The water has also been used to help in the recovery from hepatitis, indigestion, chronic inflammation of the gall bladder and related problems. The water contains added carbonation.

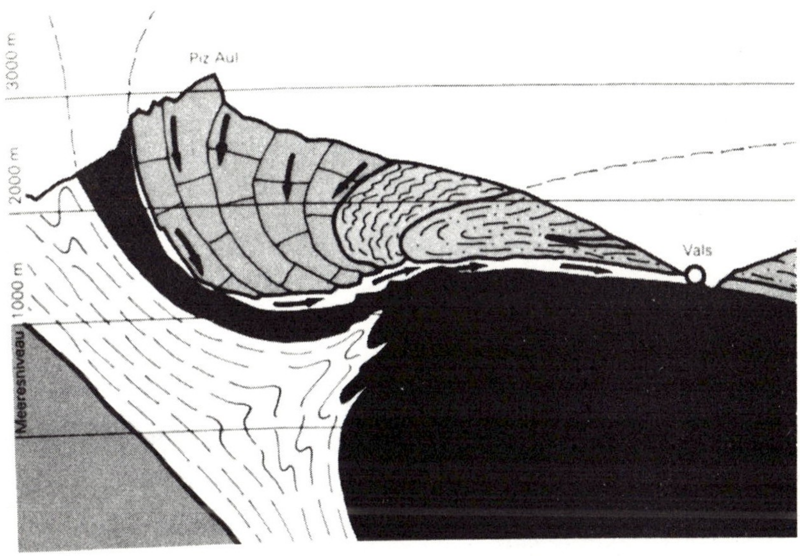

Rainwater and melted snow filters down from Piz Aul, over 3,000 meters high, to the town of Vals where it is bottled as Valser St. Petersquelle.

Valser St. Petersquelle Analysis

	Parts per Million
Calcium (Ca)	386
Magnesium (Mg)	50
Sodium (Na)	8.3
Potassium (K)	1.6
Aluminum (Al)	<.01
Bromide (Br)	.01
Chloride (Cl)	4
Copper (Cu)	.0020
Fluoride (F)	.45
Iodide (I)	.014
Lead (Pb)	<.001
Lithium (Li)	<.1
Manganese (Mn)	.03
Molybdenum	.0009
Strontium (Sr)	7.2
Vanadium (V)	.0005
Zinc (Zn)	.028
Boric Acid (H_3Bo_3)	.6
Bicarbonates (HCO_3)	345
Sulphates (SO_4)	880
Nitrates (NO_3)	.1
Phosphates (PO_4)	.02
Ammonium (NH)	<.005
Silicic Acid (H_2SiO_3)	25
Total Dissolved Solids	1710.04

The Swiss town of Vals, home of Valser St. Petersquelle.

The bottling plant of Valser St. Petersquelle.

Silber-quell

(Meltingen)

Silber-quell is rich in calcium and magnesium. Its unique taste is due to large quantities of bicarbonates which give an "Alka Seltzer" flavor. The water has a mild laxative action and stimulates the digestive tract. The Swiss take this water to aid liver function, to stimulate the intestinal tract, and to help eliminate toxins from the system. Carbon dioxide is added.

Walzhauser Mineral Water

(Walzenhausen)

Walzhauser Mineral Water originates in eastern Switzerland. The source was first discovered in 1569, and the Swiss began drinking it in the early 1800s. By 1800 the Swiss had begun to add flavors such as lemon and orange to Walzhauser. In 1909 a spa was built to attract tourists to the benefits of bathing in mineral water and to promote the sales of Walzhauser mineral water.

Nine kilometers of pipes have been constructed to carry the water from the source to the spa. The water had been transported by horse and carriage.

Doctors claim Walzhauser is helpful in treating arthritic ailments, kidney problems, and digestive problems.

Walzhauser Mineral Water Analysis

	Parts per Million
Calcium (Ca)	44.3
Magnesium (Mg)	10.3
Sodium (Na)	25
Potassium (K)	1.4
Aluminum (Al)	.008
Bromide (Br)	.006
Chloride (Cl)	3.4
Copper (Cu)	.0013
Fluoride (F)	.1
Iron (Fe)	.071
Lead (Pb)	<.0005
Lithium (Li)	<.01
Manganese (Mn)	.012
Zinc (Zn)	.031
Bicarbonates (HCO_3)	228
Sulphates (SO_4)	15
Nitrates (NO_3)	7.1
Ammonium (NH_4)	.021
Free CO_2 gas	13

Zurzacher Mineral Water

Produced in the German part of Switzerland, this water is very popular. It rises from 1,000 meters below the ground. The kinds of rock it passes through include sandstone, granite, and salt-based rocks. The water cascades 460 meters down a cliff, and flows into a pipe. It is bottled in a variety of bottle sizes, and transported by truck throughout the major portion of Switzerland. In the Italian part of Switzerland the water is transported by train.

Zurzacher is colorless, clear, and contains many natural gas bubbles. Forty-eight hours after the water comes from the ground the water still remains clear with no color distortion. As a consequence of its many natural gases, Zurzacher is extra-effervescent. The water underground has a temperature of 38.9° F but on the surface it cools further due to the evaporation of gases.

Zurzacher Water Analysis

	Parts per Million
Calcium (Ca)	11.1
Magnesium (Mg)	1.4
Sodium (Na)	290
Potassium (K)	7.9
Aluminum (Al)	.039
Bromide (Br)	.63
Chloride (Cl)	135
Copper (Cu)	.001
Fluoride (F)	.23
Iron (Fe)	.029
Lead (Pb)	.001
Lithium (Li)	1.2
Manganese (Mn)	.01
Strontium (Sr)	1
Vanadium (V)	.006
Zinc (Zn)	.003
Bicarbonates (HCO_3)	253
Sulphates (SO_4)	280
Nitrates (NO_3)	.46
Ammonium (NH_4)	.007

Mineralquelle Zurzach's immaculate bottling plant. Zurzach mineral water combines the qualities derived from pure Swiss environment and the famous Swiss precision.

YUGOSLAVIA

Radenska
Mineral Water
Water Analysis

	Parts per Million
Calcium (Ca)	227
Magnesium (Mg)	82.7
Sodium (Na)	706
Potassium (K)	93
Aluminum (Al)	.1
Bromide (Br)	.12
Chloride (Cl)	73.4
Fluoride (F)	.59
Iodide (I)	.13
Iron (Fe)	.05
Lithium (Li)	.8
Manganese (Mn)	.15
Ammonium (NH_4)	1.9
Hydrogen phosphate (HPO_4)	.1
Hydrogen carbonate (HCO_3)	2860
Nitrates (NO_3)	1.4
Sulphates (SO_4)	4.1
Carbonic Acid (CO_2)	3850
Metaboric acid (HBO_2)	4.1
Metacilicic acid (H_2SiO_3)	51.3
Total Solids	4225

Radenska

(Pomurje)

Located at the foot of the Alps in the heart of the idyllic Pomurje region, near the borders of Yugoslavia, Austria and Hungary, the Radenci Mineral Springs were first discovered in 1833 by Dr. Karl Henn. Commercial production of Radenska began in 1869. From that original spring, Radenska has grown to include 50 springs and a centrally located, modern bottling industry. In 1882, a spa was begun which has developed into a large, international health resort.

Radenska's natural sparkle comes from carbon dioxide produced in the earth's fiery interior. It works its way toward the surface through cracks and rifts, meeting and mingling with mineral-rich waters beneath Yugoslavia's Pannonian plain.

Bottled unaltered from the springs, Radenska has won a wide following, the company becoming one of the world's largest producers of mineral waters, although it was available in the United States only in the last decade. Currently, the company offers unflavored sparkling mineral water and, under the Three Hearts label, diet and regular soft drinks made with Radenska mineral water in lemon, orange, grapefruit, apple and cola flavors.

CYPRUS

Madari Mineral Water (Comet Farms Ltd.)

(Nicosia)

About twenty miles from Nicosia, in the village of Peristerona, near the Troodos Mountains, is the bottling plant of one of the most popular waters in the Arab world, Madari Mineral Water. Thirty thousand cartons containing 12 half liter bottles are sold each month in the Middle East. The water is pumped from a man-made spring.

Madari Mineral Water Analysis

	Parts per Million
Calcium (Ca)	22
Magnesium (Mg)	13
Sodium (Na)	66
Potassium (K)	1
Chloride (Cl)	5.7
Bicarbonates (HCO_3)	198
Carbonates (CO_3)	0
Sulphates (SO_4)	2.7
Total Solids at 180° C	395
Total Hardness, as $CaCO_3$	110

Moutoullas Mineral Table Water

(Nicosia)

Matoullas Mineral Table Water Analysis

	Parts per Million
Calcium (Ca)	16
Magnesium (Mg)	21
Sodium (Na)	15
Potassium (K)	.8
Chloride (Cl)	14
Bicarbonates (HCO_3)	0
Carbonates (CO_3)	0
Sulphates (SO_4)	18
Nitrites (NO_2)	0
Nitrates (NO_3)	0
Total Hardness, as $CaCO_3$	125

Moutoullas water comes from the village of the same name located on the highest mountain of the Troodos in southern Cyprus. This spring water has been bottled since 1960. It is available in both 67 centiliter glass bottles and 1.5 liter plastic bottles. It is the most popular water in Cyprus and can be found throughout the Middle East. Distributed under the trade names "O. Vivan" and "Wafa," the water is recommended for urinary tract disorders as well as arthritis, constipation, and obesity.

Glossary

alkaline —a quality of various solutions, called bases, that neutralize acids to form salts, and that turn red litmus paper blue. (See Chapter 3, *Classification of Mineral Waters.*)

anion —a negatively charged ion, for example chloride ($Cl-$) or bicarbonate (HCO_3-).

antacid —an alkaline solution which combats stomach acid, such as bicarbonate of soda in water or alkaline mineral waters.

aperient water —purgative, or laxative, water.

artesian well —a well in which water rises under pressure from a permeable stratum of rock overlaid by impermeable rock (see Chapter 3).

bicarbonate —a salt of carbonic acid.

calcium carbonate equivalent —a standard used in evaluating water hardness which may actually be caused by a variety of compounds, such as carbonates, bicarbonates, sulphates, chlorides or nitrates of calcium and magnesium. The calcium carbonate equivalent is used so hardness due to various minerals can be expressed in chemically equivalent terms, and compared from one water to another.

carbonation —saturation with carbon dioxide. Under pressure, the gas in water becomes carbonic acid.

carcinogen —any substance which tends to produce cancer.

catalyst —a substance that causes or accelerates a chemical change.

cation —a positively charged ion, such as calcium ($Ca++$) or sodium ($Na+$).

cation exchange water softner —a more specific name for the mechanical or ion exchange water softener. The ions exchanged are cations.

chalybeate water —water containing salts of iron.

charcoal filter —device for removing dissolved gases, such as chlorine, from purified water.

chemical water softner (precipitating) —a substance added to water that removes dissolved calcium and magnesium salts by forming a precipitate, or undissolved solid. Municipal water treatment plants may use this method of alleviating hardness. The precipitate is then removed by filtration. In home laundry, washing soda and similar products perform this water softening function. The drawback to home use is that there is no ready way to remove the precipitate, which can attach itself to fabrics in the wash.

Examples of this kind of softener are Climalene, Mel-O, Oakite, Perfex, sal soda and washing soda. Soap by itself removes calcium and magnesium salts from water, but a sticky precipitate is left behind.

chemical water softener (non-precipitating) —a substance added to water that softens by preventing calcium and magnesium ions from reacting with soap, thereby preventing the formation of soap curd. This process is based on the softener's ability to sequester calcium and magnesium ions; substances with this ability, most of which are in the polyphosphate family, are Calgon and Spring Rain. This kind of water softener is often incorporated into formulae for laundry soaps and detergents.

chloride of lime —a white powder used for disinfecting, which releases chlorine gas when dissolved in water.

chlorine —a tremendously active gaseous element which has been used to kill bacteria in water since 1909.

chloroform —also called trichloromethane, is a colorless, volatile, and possibly carcinogenic liquid which may appear in water supplies treated with chlorine.

cholera —a devastating disease of the digestive tract, characterized by diarrhea, vomiting, and cramps. Often fatal.

conditioned water —a term used to describe the product of various water treatments or combinations of water treatments. Frequently used to refer to softened water.

deionization —process of reducing water to a non-mineral state by passing it over a bed of resins.

deionized water —water from which both anions and cations have been removed by an ion exchange process. Only those substances which ionize in water are removed by deionization. Generally, deionized water is considered of higher quality than distilled water and is more economical to produce.

demineralized water — same as deionized water.

dissolved solids — solid materials, such as sugar or salt, which are dissolved. The materials may be ionized or non-ionized. Ionized solids can be removed from the water by ion exchange.

distilled water — water which has been purified by passing through an evaporation-condensation cycle. It contains small quantities of dissolved solids. Multiple distilling will further lower the percentage of dissolved solids.

diuretic — a substance that causes a body to release excess fluids.

ferruginous water — water containing salts of iron.

filtration — a process, either naturally occurring or artificial, whereby water passes through filters and is depleted of certain minerals or elements.

fluoridated water — water that has fluoride added for the purpose of preventing tooth decay.

fluoride — a compound of fluorine, such as stannous fluoride, used in preventive dentistry.

grain — a measure of weight derived from the average weight of a dry grain of wheat. One grain weighs about 1/7000 of a pound. An aspirin tablet weighs about 5 grains.

ground water — water contained in saturated zones of the earth. This is the supply tapped by wells and is the water source of springs.

hard water — (See *temporary hardness* and *permanent hardness*.) Water rich in calcium and magnesium salts, which causes soap to form curds. According to EPA studies, hard waters suitable for drinking may result in lower incidence of heart disease. (See also *grains per gallon* and *calcium carbonate equivalent*.)

ion — a particle with either positive or negative charge or charges. It can be made up of one element or a group of elements, for example, the calcium (Ca^{++}) or bicarbonate (HCO_3^-) ions.

ion exchange water softener — another name for mechanical water softener.

ionization — a process whereby ions are separated and/or exchanged.

lime scale — the calcium carbonate precipitate which forms when temporary hard water is heated. Lime scale accumulates inside water heaters, hot water pipes, water using appliances and cooking utensils. It has been estimated that 1/16 inch of lime scale on heating equipment pipes will reduce heating efficiency by 16% and raise fuel costs proportionately.

mechanical water softener —a device for softening water in which the hardness ions (calcium and magnesium) are exchanged for another kind of ion, usually sodium. The water that comes from the softener still contains dissolved minerals, but there are no calcium or magnesium ions present. Mechanical water softeners are also called ion exchange, cation exchange, or "zeolite" water softeners.

membrane filter —very fine filter used in filtration systems, especially in reverse osmosis.

meteoritic water —moisture precipitated from the atmosphere, such as rain, snow and hail.

milligrams per liter (mg/l) —a metric measure used in scientific and medical reports. It is approximately equal to parts per million (ppm).

mineral water —water that has large quantities of minerals, naturally collected by passing through various layers of earth and rocks to the well or spring.

municipal water —water supplied by a city for public use; tap water. Usually held in a reservoir.

naturally soft water —ground, surface, or rain water sufficiently free of calcium and magnesium salts so that no curd will form when soap is used.

naturally sparkling —naturally carbonated.

ozonization —a method of sanitizing water by using ozone which is made by forcing compressed air through a high voltage arc into the water.

permanent hardness —hardness due to sulphates, chlorides, and/or nitrates of calcium and/or magnesium. This hardness cannot be removed by heating as can temporary hardness.

pH —a number denoting alkalinity or acidity. Numbers below 7.0 indicate acidity which increases as numbers become smaller. Numbers above 7.0 indicate alkalinity which increases as numbers become larger. The pH scale runs from 0 to 14; the neutral point is 7.

The pH scale is logarithmic, that is, each whole number as the scale moves away from 7.0 is ten times more acid or more alkaline than the previous one. A pH of 5.0 is ten times more acid than pH 6.0, and pH 4.0 is ten times more acid than a 5.0 pH and 100 times more acid than a 6.0 pH. The values above 7.0 become more alkaline by the same logarithmic progression.

phenol —a poisonous and carcinogenic chemical often found in municipal water supplies. Also known as carbolic acid, phenol is a benzene derivative.

potable water —drinkable water.

ppm —abbreviation for parts per million. Approximately equivalent to mg/liter.

pure water —several conditions of water are referred to as "pure." Water may be chemically pure, bacterialogically pure, or physically pure of suspended matter. Water may, for example, be free of bacteria and yet contain pyrogens, fever-forming organisms which are still not understood completely.

The United States Pharmacopoeia's most stringent definition for pure water is water for injection, which is prepared by pharmaceutical companies. It is sold in very small bottles at high prices. This water, made under controlled laboratory conditions, is non-pyrogenic. A sample is injected into a group of live rabbits whose temperature is monitored for several days; if their temperature rises, the water is rejected.

Pure water may be prepared by ion exchange and ultra filtration sometimes followed by repeated distillation. Such water is "hungry" and will even dissolve measurable amounts of silica from glass containers.

The purest *natural* form of water would be water which is held in the clouds, just before it falls to earth as rain. As the water falls through the earth's atmosphere it collects gases and dust particles. When the water penetrates the earth, it then collects minerals and other elements.

purgative water —water that has a laxative effect.

radiological quality —the amount of radioactive materials a substance possesses.

recycled waste water —waste and sewage water which has been filtered and processed for reuse.

refined water —water processed by a combination of two or more of the following methods: cation exchange, filtration or absorption. The result is water with a hardness of less than ½ grain per gallon, with iron and heavy metal content of less than 0.1 parts per million, and free from turbidity, color, flavors and odors.

reverse osmosis —process by which water is reduced to a non-mineral state by passing through a plastic membrane under pressure, which separates the water from other elements.

soda water —a sparkling water made effervescent by a carbon dioxide charge, often containing salts giving it an alkaline, or "soda," flavor.

soft water —water sufficiently free of calcium and magnesium salts so that no curd will form when soap is used. Curd formation becomes evident when ½ grain of calcium and/or magnesium salts per gallon of water are present.

In general usage, soft water is a relative term. The Water Conditioning Foundation standards consider 1 grain of hardness or less as soft. Municipal water treatment plants consider softening to 5 or 6 grains as the optimum hardness reduction which can be economically obtained. And geological maps of natural water supplies record 3½ grains per gallon or less as soft.

Soft water may not be healthful for drinking as it does not contain certain beneficial minerals.

softened water —water which was originally hard, but from which enough calcium and magnesium ions have been removed so that no curd will form when soap is used. High in sodium chloride.

spa —a resort visited by those who wish to recuperate or revitalize their health. Spas are often built around a mineral or thermal spring.

spring water —water which flows naturally from an underground spring without the benefit of drilling or pumps.

still water —flat, uncarbonated water.

temporary hardness —hardness, caused by bicarbonates of calcium and/or magnesium, which is reduced by heating the water. When heated, the bicarbonate ion is converted to a carbonate ion, the carbonate ion reacts with calcium and magnesium ions and precipitates as insoluble calcium and magnesium salts. The precipitate is deposited as scale in hot water heaters, hot water pipes, and cooking utensils.

thermal water —hot water rising from deep within the earth where heated rocks are present.

toxin —organic poison found in living or dead organisms.

trace —less than 0.01 ppm.

trichloromethane —see chloroform.

turbidity —suspended undissolved materials in water, such as finely divided particles of sand, clay or microscopic organisms, which give color or cloudiness to the water. Turbidity can be removed by filtration.

typhoid fever —a disease characterized by enlarged intestines and ulcers, due to the typhoid bacillus. Many cases are a result of consuming infected water.

ultraviolet ray —invisible rays beyond the violet end of the visible spectrum.

water sample —to prepare a home water sample for laboratory analysis, follow this procedure: allow untreated water to run from a faucet for several minutes; use a clean bottle that will hold at least one pint and rinse the bottle several times with the water to be tested before filling it; rinse a non-metallic lid well and cover the container tightly; give the water sample to a person or organization qualified to test it. Results of the test will be stated in parts per million or grains per gallon.

water treatment —process intended to achieve one or more of the following: alleviate deficiencies such as color, turbidity, off flavors and odors; eliminate health hazards such as toxic chemicals and disease causing microorganisms and viruses; make water more satisfactory for specialized purposes.

zeolite water softener —another name for an ion exchange, cation exchange, or mechanical water softener. Derived from the name of the mineral zeolite which was commonly used in the past as the ion exchange material in mechanical water softeners.

Major U.S. Bottled Water Companies

ALABAMA

Appalachian Water Company
1901 Eleventh Avenue S
Birmingham, AL 35205
(205) 933-6333

ALASKA

Alaska's Best Water Products
5400 A Street
Anchorage, AK 99518
(907) 561-2207

Pioneer Wells
P.O. Box 914
Fairbanks, AK 99701
(907) 452-3002

Wetco
1614 Post Road
Anchorage, AK 99501
(907) 276-4115

ARIZONA

Aztec Bottled Water Company
2021 Industrial Blvd.
Lake Havasu City, AZ 86403
(602) 855-1958

Best Water
126 South Country Club Drive
Mesa, AZ 85202
(602) 964-7177

Rainbow Water Company
327 East Eighth Street
Tucson, AZ 85705
(602) 622-4646

Riviera Sun Water Vendors
943 East Marina Blvd.
Riviera, AZ 86442
(602) 758-7224

Tonopah Bottled Water
120 South 23rd Street
Phoenix, AZ 85034
(602) 244-0376

ARKANSAS

Ar-Senic Spring Water
307 Mount Ida
Hot Springs, AR 71903
(501) 623-1722

Caddo Valley Spring Inc.
P.O. Box 285
Hot Springs, AR 71903
(501) 624-1315

Diamond Water Company
179 Cedar Street
Hot Springs, AR 71901
(501) 623-1251

Mountain Valley Spring Water
150 Central Avenue
Hot Springs, AR 71901
(501) 623-6671

Pinnacle Mountain Waters
775 South Government Street
Fayetteville, AR 72701
(501) 442-6074

Pyramid Spring Water Company
P.O. Box 23
St. Joe, AR 72675
(501) 741-2461

CALIFORNIA

A-1 Seltzer & Beverage Co.
6878 Beck Avenue
No. Hollywood, CA 91605
(213) 875-1975

Alhambra National Water
2450 Washington Avenue
San Leandro, CA 94577
(415) 352-0910

American Natural Springs
P.O. Box 481
Palo Alto, CA 94301
(415) 321-6000

Applegate Spring Water Company
P.O. Box 1
Applegate, CA 95703
(916) 878-7089

Aqua Pure Drinking Waters
12322 Dahlia Avenue
El Monte, CA 91732
(818) 350-5711

Arctic Pure Water
1250 South Wilson Way
Stockton, CA 95205
(209) 941-8558

Arrowhead Waters
601 East Potrero Grande Drive
Monterey Park, CA 91754
(818) 705-3338

Artesian Water
1806 Market Street
San Francisco, CA 94102
(415) 621-1515

Bartlett Mineral Springs
1686 Union Street
San Francisco, CA 94123
(415) 474-5210

Bastanchury Water Company
601 West Valencia Drive
Fullerton, CA 92632
(714) 870-4022

Black Mountain Springs
800 Alameda De Las Pulgas
San Carlos, CA 94070
(415) 221-8570

Calistoga Mineral Water Co.
865 Silverado Trail
P.O. Box 97
Calistoga, CA 94515
(707) 942-4100

Cole's Well Water
1604 Wilson Road
Ramona, CA 92065
(619) 789-1700

Crystal Geyser Water Company
501 Washington Street
P.O. Box 304
Calistoga, CA 94515
(707) 942-0500

Crystal Springs Water Co.
P.O. Box 3786
Eureka, CA 95501
(707) 443-7171

Crystal Springs Water Company
1111 La Madrona Drive
Santa Cruz, CA 95506
(408) 429-1418

Crystal Springs Water Company
3215 Rockview Place
San Luis Obispo, CA 93401
(805) 543-5063

Elton's Spring Water
7210 Jordan Avenue, Ste. D114
Canoga Park, CA 91303
(818) 709-7344

Forest Lake Products Inc.
P.O. Box 239
Cobb, CA 95426
(707) 928-5347

Gladstone Springs Bottled Water
29208 Posey Way
Rancho Palos Verde, CA 90274
(213) 541-9259

High Sierra Springs
11916 Woodley Avenue
Granada Hill, CA 91344
(818) 709-9344

Hinckley & Schmitt
4414 North York Blvd.
Los Angeles, CA 90041
(213) 255-0106

Indian Head Water Company
226 South Avenue 54
Los Angeles, CA 90042
(213) 256-2265

CALIFORNIA (continued)

Knoxage Water
227 Ninth Avenue
San Diego, CA 92101
(619) 234-3333

Matilija Water Company
164 West Park Row Avenue
Ventura, CA 93001
(805) 643-4675

McKesson Water Division
P.O. Box 41049
Los Angeles, CA 90041
(213) 259-2000

Mendocino Mineral Water
P.O. Box 1400
Mendocino, CA 95460
(707) 937-4597

Mile High Mountain Spring Water
1269 Linda Vista Drive
San Marcos, CA 92069
(619) 744-9542

Modesto Bottled Water Company
P.O. Box 4643
Modesto, CA 95352
(209) 527-6764

Mountain Fresh Water Company
1200 Arden Way
Sacramento, CA 95815
(916) 927-5567

Mountain Springs Water Company
895 Innes Avenue
San Francisco, CA 94124
(415) 821-3444

Napa Valley Spring Water Co.
1506 Grant Street
P.O. Box 797
Calistoga, CA 94515
(707) 942-6482

Niagara Drinking Water
17842 Cowan Street
Irvine, CA 92714
(714) 863-1400

Oasis Drinking Waters
1506 North Clinton Street
Santa Ana, CA 92703
(714) 554-6000

Old San Francisco Seltzer
5813 Uplander Way
Culver City, CA 90230
(213) 443-8808

Palomar Spring Water
1270 West Mission Avenue
Escondido, CA 92107
(619) 489-5787

Polar Water Company
1029 West Fremont Street
Stockton, CA 95203
(209) 466-6766

Pure Flo Water Company
7737 Mission Gorge Road
Santee, CA 92071
(619) 448-5120

Puritan Water Company
1080 South Cypress, #D
La Habra, CA 90631
(213) 691-6739

Ramona Bottling Inc.
P.O. Box 1195
Ramona, CA 92065
(619) 789-0174

Rocky Mountain Water
1880 Chris Lane
Anaheim, CA 92805
(714) 634-1101

Rodgers Bottling Company
1243 Florida Street
Vallejo, CA 94590
(707) 644-4151

Sierra Spring Water Company
1801 R Street
Sacramento, CA 95814
(916) 446-5341

Sparkletts Drinking Water
4500 York Blvd.
Los Angeles, CA 90041
(213) 259-2237

Triple AAA Water Company
P.O. Box 1379
Brawley, CA 92227
(619) 344-2075

Vichy Springs Mineral Water
2605 Vichy Springs Road
Ukiah, CA 95482
(707) 462-9515

Vittel Imports USA, Inc.
1686 Union Street, #305
San Francisco, CA 94123
(415) 474-1252

Wallace Water Company
524 South Ninth Street
Modesto, CA 95351
(209) 578-0609

Yosemite Waters
1120 South Parallel Avenue
Fresno, CA 93702
(209) 268-8469

COLORADO

Clark Spring Water Company
319 Clark Street
Pueblo, CO 81003
(303) 543-1594

Colorado Mountain Water
1055 County Road, #220
Durango, CO 81301
(303) 259-5300

De Loach's Water Inc.
P.O. Box 906
Lamar, CO 81052
(303) 336-5201

Deep Rock Water South
2930 Janitell Road
Colorado Springs, CO 80906
(303) 579-6550

Eldorado Artesian Springs
P.O. Box 445
Eldorado Springs, CO 80025
(303) 499-1316

Ev'r Clear Artesian Water
4120 Globeville Road
Denver, CO 80216
(303) 294-0311

Rifle Ice/Mountain
128 West Avenue
Rifle, CO 81650
(303) 625-1701

Silver Lion Coors Inc.
121 Apollo Road
Box 7
Montrose, CO 81402
(303) 249-5533

CONNECTICUT

Crystal Rock Spring Water
313 Long Ridge Road
Stamford, CT 06902
(203) 325-1357

Evian Waters of France
600 West Putnam Avenue
Greenwich, CT 06830
(203) 629-3642

Foxon Park Beverages
103 Foxon Blvd.
East Haven, CT 06513
(203) 467-7874

Granite Springs Beverages
100 Smith Road
Thomaston, CT 06787
(203) 283-4546

Great Bear Spring Company
255 James Street
New Haven, CT 06473
(203) 777-2534

Guida-Seibert Dairy
433 Park Street
New Britain, CT 06051
(203) 743-3410

CONNECTICUT (continued)

Light Rock Spring Water
Nine Balmforth Avenue
Danbury, CT 06810
(203) 743-3410

Manitock Spring Water Company
343 Boston Post Road
Box 297
Waterford, CT 06385
(203) 443-5000

Mount Claire Spring Water
160 Perkins Street
Torrington, CT 06790
(203) 489-3804

Mountain Spring Water
1286 East Main Street
Meridan, CT 06450
(203) 237-3500

Pequot Spring Water Company
Spring Street Extension
Glastonbury, CT 06033
(203) 633-9467

Perrier Group
777 West Putnam Avenue
Greenwich, CT 06830
(203) 531-4100

Poland Spring Corporation
777 West Putnam Avenue
Greenwich, CT 06830
(203) 629-2360

Sand Springs Spring Water
1120 Boston Post Road
Darien, CT 06820
(203) 655-1122

Triple Springs Water Company
199 Ives Avenue
Meriden, CT 06450
(203) 235-8374

DELAWARE

Mr. Natural Bottled Water
P.O. Box 490
Ocean View, DE 19970
(302) 539-8900

DISTRICT OF COLUMBIA

Giant Food Inc.
P.O. Box 1804
Washington, D.C. 20013
(301) 341-3970

FLORIDA

Blue Crystal Water Company
4120 Roush Avenue
Orlando, FL 32803
(305) 898-2000

Carolina Mountain Spring Water
6604 Miami Lakes Drive E.
Miami Lakes, FL 33014
(305) 822-6502

Crystal Water Company
6731 Philips Pkwy Drive N
Jacksonville, FL 32224
(904) 268-4199

Kissimmee Spring Water Company
15 S. Airport Rd.
Box 2888
Kissimmee, FL 32741
(305) 847-2617

Perfection Water Company
629 Airport Road
Naples, FL 33940
(813) 643-0466

Ramsay Water Company
45 N.E. 96 Street
Miami, FL 33138

Southeast-Atlantic Corp.
5900 N.W. 72 Ave.
Box 520365
Miami, FL 33152
(305) 887-6535

Sun Seltzer Inc.
2923 N.W. 17 Terrace
Ft. Lauderdale, FL 33311
(305) 739-4540

Syfo Water Company
12955 Northeast 14 Avenue
North Miami, FL 33161
(305) 891-2200

White Springs Water Company
8774 Fourth Avenue
Box 9238
Jacksonville, FL 32208
(904) 764-0511

Wilson Bottling Corporation
3550 Northwest 110 Street
Miami, FL 33167
(305) 685-5061

Zephyrhills Bottled Water
1850 Fort King Road
Zephyrhills, FL 34248
(813) 788-0466

GEORGIA

Crystal Springs Pure Water
3620 Waldrop Road
Decatur, GA 30034
(404) 241-8144

Georgia Mountain Water
3929 Highway 5
Marietta, GA 30061
(404) 928-9971

Lithia Springs Water Company
2910 Bankhead Hwy.
Box 713
Lithia Springs, GA 30057
(404) 944-3880

Stone Mountain Water Company
P.O. Box 1687
Stone Mountain, GA 30086

Windsor Spring Water Company
Route 5
P.O. Box 214
Hephzibah, GA 38015
(404) 798-2373

HAWAII

Hawaiian Water Partners
3615 Hardin Ave., Ste. 408
Honolulu, HI 96816
(808) 733-1000

Menehune Water Company
1095 Dillingham Blvd. I-6
Honolulu, HI 96817
(808) 841-3355

The Water Man
774 Lower Main Street
Wailuku, Maui, HI 96793
(808) 242-6847

IDAHO

Blue Springs Water Company
4806 Oberland Road
Boise, ID 83705
(208) 336-3949

Spring of Youth
1916 Greenhurst
Nampa, ID 83651
(208) 466-2842

ILLINOIS

Alpine Valley Waters
16900 South Lathrop
Harvey, IL 60426
(312) 757-5835

Aquarius Water
P.O. Box 350
Bloomingdale, IL 60108
(312) 364-4050

D'Angelo Natural Spring
Routes 72 & 68
Dundee, IL 60118
(312) 426-4025

Peoria Mineral Springs
701 W. Martin Luther King
Peoria, IL 61605
(309) 676-7951

Quality/Absopure Water Company
1 East Tremont
Champaign, IL 61820
(217) 398-5011

Sparkling Spring Mineral Water
1621 Shields
Chicago Heights, IL 60411
(312) 755-7727

ILLINOIS (continued)

Sparkling Spring Mineral Water
202 West Main Street
Plano, IL 60545
(312) 552-4124

INDIANA

Greenwood Springs Water
611 West State Street
Fort Wayne, IN 46808
(219) 426-2049

Kline's Quality Water Inc.
2250 North Ritter Avenue
Indianapolis, IN 46218
(317) 353-9444

Pri-Pak Inc.
P.O. Box 450
Sunman, IN 47041
(812) 623-3300

Spring Water Inc.
P.O. Box 535
Cayuga, IN 47928

Wyandotte Spring Corporation
P.O. Box 166
New Salisbury, IN 47161
(812) 364-4124

IOWA

Dasco Inc.
Highway 61
Box 357
Mediapolis, IA 52637
(319) 394-3907

Humboldt Distilled Water
610 Thirteenth Street N
Humboldt, IA 50548
(515) 332-1602

Magic Mountain Bottled Water
3836 Eastern Avenue
Davenport, IA 52807
(319) 391-4414

Stonebridge Company
1700 East Broadway
Des Moines, IA 50313
(515) 262-5145

KANSAS

Conway Springs Water
P.O. Box 977
Wichita, KS 67201
(316) 522-3131

Mineral Rock Springs
1015 Metropolitan Avenue
Kansas City, KS 66100
(913) 321-5557

KENTUCKY

Anita Spring Water Company
3600 Chamberlain Lane, #222
Louisville, KY 40222
(502) 423-0485

Foster Water Company
6200 Old Heady Road
Jeffersontown, KY 40299
(502) 897-3113

Kentucky Bubbling Springs
P.O. Box 7155
Louisville, KY 40207
(502) 897-3113

Mitchell Water Delivery
Box 191
Martin Hill
Shepperdsville, KY 40165
(502) 366-8752

Samuels Spring Water Company
P.O. Box 8
Deatsville, KY 40016

LOUISIANA

Kentwood Spring Water Inc.
3418 Howard Ave.
Box 52043
New Orleans, LA 70152
(504) 821-4505

Music Mountain Water Company
1512½ Louisville Avenue
Monroe, LA 71201
(318) 323-6525

MAINE

Nezinscot Spring Bottling
RFD 1, Box 280
Turner, ME 04282
(207) 255-3222

MARYLAND

All Pure Spring Water Inc.
Route 4, Box 50A
Deer Park, MD 21550
(301) 334-2124

Appalachian Water Inc.
P.O. Drawer Aa
Deer Park, MD 21550
(301) 334-1500

Green Spring Water Company Inc.
R.R. 1, Box 65
Big Springs, MD 21722
(301) 739-5571

Kronheim Company Inc.
99 Ray Road, Box 7331
Baltimore, MD 21227
(301) 242-8000

Snow Valley Water Company
1908 West Street, Box 1949
Annapolis, MD 21401
(301) 266-6600

Talbert's Ice & Beverage
5234 River Road
Bethesda, MD 20816
(301) 652-3000

MASSACHUSETTS

A-1 Spring Water Corporation
101 Linebrook Road
Ipswich, MA 01938
(617) 599-2680

Balsam's Spring Water
6 Farrwood Drive
Andover, MA 01810

Belmont Springs Water
Country Club Lane, Box 1
Belmont, MA 02179
(617) 484-1356

Berkshire Springs Inc.
Norfolk Road
Southfield, MA 01259
(413) 229-2086

Boston Water Cooler Inc.
211 Main Street
North Reading, MA 01864
(617) 944-8880

Castle Rock Springs
509 Main Street
Saugus, MA 01906
(617) 233-0827

Fahey Beverage Company
39 Elmvale Place
Pittsfield, MA 01201
(413) 442-7313

Greylock Spring Water
P.O. Box 2020-9 Greylock
Worcester, MA 01613
(617) 752-5599

Gueli Spring Water Company
9 Bradley Avenue
Methuen, MA 01844
(617) 683-8873

Holyoke Distilled Water
39 Laurel Street
Holyoke, MA 01040
(413) 533-5098

Ionics Inc.
65 Grove Street
Watertown, MA
(617) 926-2500

Manhattan Bottling Co., Inc.
135 Potter Street
Box A2060
New Bedford, MA 02741
(617) 993-9916

MASSACHUSETTS (continued)

Miscoe Spring Beverage
89 Northbridge Road
Mendo, MA 01756
(617) 473-0550

New England Spring Water
P.O. Box 179
Boston, MA 02118
(617) 353-0684

Ravenwood Spring Water Company
13 Beauport Avenue
Gloucester, MA 01930
(617) 283-1778

Refreshment Connection
351 Walnut Street Ext.
Agawam, MA 01001
(413) 786-3407

Simpson Spring Company
719 Washington Street
South Easton, MA 02375
(617) 238-2741

Water Marketers Inc.
251 Beacon Street
Boston, MA 02116
(617) 267-7477

Webster Bottling Company
58 E. Main Street
Box 575
Webster, MA 01570
(617) 943-0217

MICHIGAN

Arbor Springs Water Company
1440 Plymouth Road
Ann Arbor, MI 48105
(313) 668-8270

Harbor Spring Water Company
P.O. Box 104
Harbor Springs, MI 49740

Shay Water Company
320 West Bristol
Saginaw, MI 48602
(517) 755-3221

MINNESOTA

Chippewa Springs Corporation
122 Southeast Eighth Street
Minneapolis, MN 55414
(612) 379-4141

Cold Spring Water Company
219 North Red River Avenue
Cold Spring, MN 56320
(612) 685-8686

Kandiyohi Bottled Water
Route 4
Willmar, MN 56201
(612) 235-7194

Viking Waters Inc.
P.O. Box 212
Glenwood, MN 56334
(612) 634-4328

MISSISSIPPI

Dixie Springs Spring Water
P.O. Box 727
Summit, MS 39666
(601) 276-7666

Mississippi Bottled Water
P.O. Box 5393
Jackson, MS 39216
(601) 981-1231

MISSOURI

Chouteau Artesian Water
P.O. Box 391
Sedalia, MO 65301
(816) 826-8144

Mineral Water System
201 East Broadway
Excelsior Springs, MO 64024
(816) 637-0752

Pepsi-Cola Bottling Company
Highway 10 East
Excelsior Springs, MO 64024
(816) 826-8144

Springfield Water Inc.
1921 East McDaniel Street
Springfield, MO 65802
(417) 862-3729

Swiss Altima
1402 Mercantile Tower
Kansas City, MO 64106
(816) 763-1953

Tawanee Water Dist. Co.
1428 North Broadway
St. Louis, MO 63102
(314) 621-2280

Tyler Mountain Water Company
7359 Hazelwood Blvd.
Hazelwood, MO 63042
(314) 522-1043

MONTANA

Crown Beverage Company
P.O. Box 8959
Missoula, MT 59807
(406) 728-8100

Lewistown Coca-Cola Bottling
204 First Avenue North
Lewiston, MT 59457
(406) 538-3991

Yellowstone Water Service
2109 First Avenue North
Billings, MT 59101
(406) 259-9119

NEBRASKA

American Bottled Water Company
8934 L Street
Omaha, NE 68127
(402) 339-5453

Deep Rock Fontenelle
1419 S. Saddle Creek Road
Omaha, NE 68127
(402) 558-5116

NEVADA

Crystal Springs Water Company
901 South Center Street
Reno, NV 89504
(702) 323-4710

NEW HAMPSHIRE

Honey Book Mountain Spring
P.O. Box 1
Route 9A
Spofford, NH 03462
(603) 363-4562

Intervale Spring Water
P.O. Box 597
Wilton, NH 03086
(603) 654-2728

NEW JERSEY

American Beverage Company
281 Passaic Street
Garfield, NJ 07026
(201) 473-6463

Belmar Spring Water Company
410 Grove Street
Ridgewood, NJ 07450
(201) 444-1010

Betta Water Company
905 East County Line Road
Lakewood, NJ 08701
(201) 472-6900

Brookdale Beverage Company
955 Bloomfield Avenue
Clifton, NJ 07012
(201) 472-6900

Cortco International Corp.
419 Hoboken Avenue
Jersey City, NJ 07306
(201) 653-0459

Electrified Water Company
101 North Park Street
East Orange, NJ 07017
(201) 678-8100

Horizon Bottled Water
450 Raritan Center
Edison, NJ 08818
(201) 225-2233

Indian Spring Water Company
350 Route 46
Rockaway, NJ 07866
(201) 627-0379

NEW JERSEY (continued)

Johanna Farms Inc.
P.O. Box 272
Flemington, NJ 08822
(202) 782-4515

Kepwel Spring Water Company
100 Cold Indian Spring Road
Ocean, NJ 07712
(201) 922-1030

Krauszers Dairy Inc.
National Road
Edison, NJ 08818
(201) 287-2800

Modern Beverages Inc.
P.O. Box 492
Blairstown, NJ 07078
(201) 362-8847

Mullins Spring Water Company
675 New Brunswick Avenue
Perth Amboy, NJ 08861
(201) 442-1337

National Spring Water Company
419 Southside Avenue
Haledon, NJ 07508
(201) 790-6600

Ozone Water Company
120 Main Street
Orange, NJ 07050
(201) 674-3244

Portasoft Company
1122 South Avenue
Westfield, NJ 07090
(201) 233-4300

Shamrock Water Company
40 Blaine Street
Box 27
Passaic, NJ 07055
(201) 778-6474

NEW MEXICO

Cascade Bottled Water Company
100 South Lake Street
Farmington, NM 87401
(505) 325-1859

Southeastern Public Service
601 Commercial Avenue N.E.
Albuquerque, NM 87102
(505) 243-7733

NEW YORK

Chemung Spring Water Company
P.O. Box 241
Chemung, NY 14825
(607) 529-3262

Cott Beverage Distributor
530 Homestead Avenue
Mt. Vernon, NY 10550
(914) 668-0561

Duxbury Vermont Springs
1056 Washington Avenue
Bronx, NY 10456
(212) 993-0200

Excelsior Spring Water
454 Broadway
Saratoga Springs, NY 12866
(518) 583-00554

Filtered Water Service
419-421 West 55 Street
New York, NY 10019
(212) 582-8888

Gramatan Springs Company
125 Midland Avenue
Bronxville, NY 10708
(914) 337-3319

Great Bear Spring Company
2 Warehouse Lane
Elmsford, NY 10523
(914) 592-4353

Health Waters Inc.
P.O. Box 1061
West Babylon, NY 11704
(516) 587-6172

Joy Springs Inc.
621 Deerfield Drive
Oneida, NY 13421

Leisure Time Spring Water
P.O. Box K, Route 42
Kiamesha Lake, NY 12751
(914) 794-7040

Mt. Laurel Spring Water
111 Broadway
New York, NY 10006
(212) 349-3030

Nehi Beverage Inc.
P.O. Box 4147
Albany, NY 12204
(518) 434-1191

North Country Natural Spg. Water
P.O. Box 123
Port Kent, NY 12975
(518) 834-4644

Risley Cold Springs
R.D. 1, Box 22
Whitesboro, NY 13492
(315) 736-1226

Saratoga Mineral Water
11 Geyser Road
Saratoga Springs, NY 12866
(518) 584-6363

Stanford Springs Inc.
Cold Spring Road
Standfordville, NY 12581
(914) 868-7600

NORTH CAROLINA

Arcadia Farms Inc.
P.O. Box 631
Arden, NC 28704
(704) 684-3556

Carolina Mountain Spring Water
P.O. Box 407
Cashiers, NC 28717
(704) 743-2129

Midas Spring Water Inc.
P.O. Box 16143
Charlotte, NC 28216
(704) 743-2129

Milko Inc.
P.O. Box 16160
Asheville, NC 28816
(704) 254-9560

The Water House
800 Chatham Road
Winston-Salem, NC 27101
(919) 761-0884

Triton Water Company
739 S. Worth Street
Box 2690
Burlington, NC 27215
(919) 227-8838

NORTH DAKOTA

Granite Spring Water Company
2100 Fourth Avenue N.W.
Minot, ND 58701
(701) 838-3910

OHIO

American Beverage Company
1023 Stiemmel Road
Columbus, OH 43223
(614) 443-4515

Anita Springs Bottled Water
3303 Felicity Drive
Cincinnati, OH 45211
(513) 481-7873

Aquarius Spring Water Co.
49 South Worthington Street
Youngstown, OH 44501
(216) 744-2121

Cherry Knoll Spring Water
7130 Leavitt Road
Amherst, OH 44001
(216) 986-2197

Collingwood Water Company
710 Berdan
Toledo, OH 43610
(419) 241-6715

Distillata Company
1542 East 24 Street
Cleveland, OH 44114
(216) 771-2900

OHIO (continued)

Geauga Springs Inc.
10729 Kinsman Road, Rt. 87
Newbury, OH 44065
(216) 338-5003

Highland Springs
P.O. Box 278
Rayland, OH 43943
(614) 859-4734

Howland Springs Water Company
8707 Howland Springs S.E.
Warren, OH 44484
(216) 856-2404

Magnetic Springs Water Co.
1801 Lone Eagle Street
Columbus, OH 43228
(614) 870-7770

Mayfield Importing Company
3820 Lakeside
Cleveland, OH 44114
(216) 431-3344

Minnehaha Water Company
3700 Prospect Avenue
Cleveland, OH 44115
(216) 431-0243

National Ice House Water
100 East Fifth
Mansfield, OH 44901
(419) 524-1312

Polar Water Company
3280 East Eleventh Avenue
Columbus, OH 43219
(614) 252-1135

Sand Rock Mineral Water
1312 East Tuscarawas Street
Canton, OH 44707
(216) 453-4708

Shawnee Valley Water Company
750 Bell Fountain Avenue
Lima, OH 45801
(419) 229-7336

Talawanda Springs Inc.
3965 Rosslyn Drive
Cincinnati, OH 45209
(513) 721-3626

Whitehouse Artesian Springs
44285 Telegraph Road
Elyria, OH 44035
(216) 322-1317

OREGON

Aqua Cool Bottled Water
802 North East Davis
Portland, OR 97232
(503) 233-9991

Bonneau Products Company
81 N.E. Columbia Blvd.
Portland, OR 97211
(503) 289-7181

Mountain Spring Water Company
5520 Upper River Road
Grants Pass, OR 97526
(503) 476-7384

PENNSYLVANIA

Andorra Spring Water
2201 Barren Hill Road
Conshohocken, PA 19428
(215) 828-6330

Aqua Penn Spring Water
511 Toftrees Avenue, #205
State College, PA 16803
(814) 231-2076

Blue Mountain Spring Water
1840 South 19 Street
Harrisburg, PA 17104
(717) 986-9283

Cloister Pure Spring Water
1060 South State Street
Ephrata, PA 17522
(717) 733-7901

Ephrata Diamond Spring
1140 South State Street
Ephrata, PA 17522
(717) 733-1264

Glen Summit Springs Water
1000 Rutter Ave., Box 1108
Forty Fort, PA 18704
(717) 287-3107

Glen Summit Springs Water
405 South Main Avenue
Scranton, PA 18504
(717) 342-4213

Herrmann's Water
833 Buckeye Drive
Sharpsville, PA 16150
(412) 962-7434

Miller Spring Water Company
P.O. Box 15
Schnecksville, PA 18103
(215) 433-6999

Monroe Bottling Company
560 North Main Street
Pittston, PA 18640
(717) 655-5497

Mt. Meadows Spring Water
P.O. Box 299
Bechtelsville, PA 19505
(215) 369-0331

Pocono Artesian Waters Co.
P.O. Box 241
Bear Creek, PA 18602
(717) 822-6907

Pocono Mountain Bottling Co.
57 West Chestnut Street
Wilkes Barre, PA 18705
(717) 822-7695

Rockwood Sales Inc.
2138 East Sergeant
Philadelphia, PA 19125
(215) 323-5343

Rosedale Water Company
6216 Verona Road
Verona, PA 15147
(412) 793-7495

Ross-Common Spring Water
620 Tenth Avenue
Bethlehem, PA 18018
(215) 867-5815

Short Mountain Spring Water Co.
R.D. #2, Box 433A
Cogan Station, PA 17728
(717) 398-7519

Silver Spring Water Company
77 Lockhart Street
Wilkes-Barre, PA 18702
(717) 823-7019

Spring Water Company
1030 Trenton Road
Fallsington, PA 19054
(215) 295-5230

Summit Springs
Route 4
Conneaut Lake, PA 16316
(814) 382-7873

Swiss-Koldrok Water Company
1817 Fifth Avenue
McKeesport, PA 15132
(412) 241-6391

RHODE ISLAND

Dybala Bottling Company
46 Thomas Street
Woonsocket, RI 02895
(401) 762-0035

Girard Spring Water Company
1100 Mineral Spring Avenue
North Providence, RI 02904
(401) 725-7298

SOUTH CAROLINA

Case Pure Water Company
P.O. Box 1568
Murrells Inlet, SC 29576

SOUTH DAKOTA

Pearl Spring Water Company
P.O. Box 598, R.R. #3
Huron, SD 57350
(605) 352-5750

TENNESSEE

Burgie Industries Inc.
90 W. Desoto Ave., Box 13345
Memphis, TN 38113
(901) 948-8694

Continental Water
1938 Lynnbrook
Memphis, TN 38116
(901) 398-2300

L'eau Beau
901 Division Street
Nashville, TN 37203
(615) 242-3737

Memphis Water Bottling Co.
4249 North Watkins
Memphis, TN 38127
(901) 358-5279

Willow Creek Water Company
6084 Apple Tree Drive
Memphis, Tn 38115
(901) 366-7576

TEXAS

5 Springs Corporation
4751 Durazno, Box 11470
El Paso, TX 79982
(915) 545-4500

Alpine Water Company
4718 McCarty Blvd.
Amarillo, TX 79110
(800) 221-1100

Ambrosia Water of Brazos
1501 Fm 2818, #303
College Station, TX 77840
(409) 846-0000

Artesia Waters Inc.
4671 Walzem Road
San Antonio, TX 78218
(512) 654-0293

Cargile Water Company
5841 South First
Abilene, TX 79605
(915) 692-5774

Desert Springs
138 Palm Drive
San Antonio, TX 78228
(512) 521-4402

DLG Sparkling Water
4531 Ayres Street
Corpus Christi, TX 78415
(512) 851-0800

Everest Drinking Water Co.
1401 S. Padre Island Drive
Corpus Christi, TX 78416
(512) 851-8946

Kalil Bottling Company
7328 Boeing Drive
El Paso, TX 79925
(915) 778-4413

Mystic Ozark Water Company
5841 South First Street
Abilene, TX 79605
(915) 692-5774

Oasis Water Company
4109 Todd Lane
Austin, TX 78744
(512) 562-2626

Renoir Sparkling Water Co.
15600 Drummet Blvd., #500
Houston, TX 77032
(713) 987-0101

Tarpley Valley Spring Water
P.O. Box 50
Tarpley, TX 78883
(512) 821-7044

Texas Spring Inc.
P.O. Box 585
Salado, TX 76571
(817) 947-5393

UTAH

Stiles Beverage Company
3507 South 700 East
Salt Lake City, UT 84106
(801) 467-9421

VERMONT

Green Mountain Springs
R.F.D. #1
Sudbury, VT 05733
(802) 773-5578

Northeast Kingdom Spring
P.O. Box 209
West Burke, VT 05871
(802) 467-8321

VIRGINIA

Baker Seawright Corporation
R.F.D. 1, Box 263
Mt. Sidney, VA 24467
(703) 866-6255

Diamond Springs Water Company
4100 Diamond Springs Drive
Richmond, VA 23231
(804) 795-2038

Miller's Spring Water Inc.
925 Cavalier, Box 6314
Chesapeake, VA 23323

Spring Grove Water Company
Highway 40, Box 35
Spring Grove, VA 23881
(804) 866-8921

Tannourine Mineral Water
P.O. Box 1195
McLean, VA 22101
(703) 827-0207

Virginia Bottled Water Co.
P.O. Box 128
Swoope, VA 24479
(703) 337-7111

WASHINGTON

Carnation Water Company
10304 296 Avenue N.E.
Carnation, WA 98014
(206) 788-5546

Cullyspring Water Company
1910 21 Avenue South
Seattle, WA 98144
(206) 329-4942

Diamond Water Company
4123 North 35 Street
Tacoma, WA 98407
(206) 752-5046

WISCONSIN

Bethesda Roxo Waters
574 Elizabeth Street
Waukesha, WI 53187
(414) 755-2316

Buffalo Dons Artesian
Hillview Road, Route 1
Mischicot, WI 54228
(414) 755-2316

Cascade Supply Corporation
647 West Virginia Street
Milwaukee, WI 53204
(414) 276-7800

Coca-Cola Bottling Company
1740 Ind'l Pkwy., Box 656
Marinette, WI 54143
(715) 735-5133

Country Crystal Inc.
6120 Tower Avenue
Superior, WI 54880
(715) 392-7479

Mineral Spring Water Company
3027 Jackson Drive
Oshkosh, Wi 54901
(414) 231-9290

Rocky Dell Springs Inc.
P.O. Box 64
Woodman, WI 53821
(608) 533-3767

WYOMING

Hillcrest Water Company
333 East H Street, Box 1586
Casper, WY 82601
(307) 237-9863

McCormick, Inc.
P.O. Box 832
Gillette, Wy 82716
(307) 682-2588

For Further Reading

Accum, Frederick. *Guide to the Chalybeate Spring of Thetford.* London: C. Green, 1819.

Adams, Ruth and Murray, Frank. *Minerals: Kill or Cure?* New York: Larchmont Books, 1974.

American Public Health Association, American Waterworks Association and Water Pollution Control Federation. *Standard Methods for the Examination of Water and Waste-Water, Including Bottom Sediments and Sludges.* 12th ed. Albany, New York: Boyd Printing Co., 1965.

American Water Works Association. "Water Purifiers for the Home: What to Tell Customers About Them." *Opflow* 3 (November 1977): 1, 7.

Anderson, Terence W., M.D. and Mackay, J. S. "Sudden Death, an Ischemic Heart Disease—Correlation with Hardness of Local Water Supply." *New England Journal of Medicine* 280 (1969): 805-807.

Baker, M. N. *The Quest for Pure Water.* New York: American Water Works Association, 1948.

Baldwin, Carol. "How to Take the Sting Out of Your Drinking Water." *Prevention Magazine* 29 (December 1977): 96.

Behrman, A. S. *Water is Everybody's Business.* Garden City, New York: Anchor Books, Doubleday & Co., Inc. 1968.

Bragg, Paul C., N.D., M.D. *The Shocking Truth About Water.* 9th ed. Desert Hot Springs: Health Science, 1976.

Budge, Sir Ernest A. T. W. *The Gods of the Egyptians.* New York: Dover Publications, 1969.

Bukro, Casey. "U.S. Studying Hazards in Tap Water." *Chicago Tribune,* 30 March 1975, p. 1.

Burgstahler, Albert W. "Dental and Medical Aspects of Fluoridated Drinking Water." *Kansas Academy of Science* 68 (1965). Nos. 2 and 3.

Burk, Delaney and Yiamouyiannis, John, Ph.D. *Fluoridation and Cancer.* Report prepared for the House Sub-Committee on Intergovernmental Relations and Human Resources. Delaware, Ohio: The National Health Federation, September 21, 1977.

Carroll, Rick. "Vending Machines, Water at 20 Cents a Pop." *San Francisco Chronicle,* 31 March 1977, Part II, p. 19.

Clark, Linda, M.D. *The Best of Linda Clark, Some Unusual Approaches to Health.* New Canaan, Connecticut: Keats Publishing Inc., 1974.

Condlon, Edward J. "Bank of Manhattan Company Continues to Get Reminders of Its Start in the Business of Selling Water." *New York Times,* 10 August 1941, Part III, p. 1.

Cotar, Charles, M.D. *A Treatise on the Mineral Waters of Vichy.* London: H. K. Lewis, 1913.

Crook, James K., A.M., M.D. *Mineral Waters of the United States and Their Therapeutic Uses.* New York: Lea Brothers, & Co., 1829.

"Drinking Water." *Wet Magazine* 1 (1976): 4-13.

Exner, F. B. "A Hoax from Harvard." *The National Health Federation Bulletin* 13 (June 1967).

Foord, Alfred Stanley. *Springs, Streams, and Spas in London.* London: T. Risher Unnin, 1910.

Fox, Martin. *Healthy Water for Longer Life.* Orleans, MA: Healthy Water Research, 1984.

Gould, Robert W., ed. *Trace Inorganics in Water.* Washington, D.C.: American Chemical Society Publications, 1968.

Green, Timothy and Green, Maureen. *The Good Water Guide.* London: Rosendale Press, 1985.

Harker, Timothy L. "Regulation of Drinking Water, Legal and Policy Considerations." *American Bottled Water Reporter* (August 1977): 6, 8.

Hewett, Donnel Foster and Crecking, G. W. *The Warm Springs of Georgia, Their Geologic Relations and Origin.* U.S. Geological Survey Water Supply Report. Washington, D.C.: Government Printing Office, 1937.

James, E. O. and Mercer, F. A. B. "Water, Water Gods." *Encyclopedia of Religion and Ethics.* New York: Charles Scribner and Sons, 1922.

Kersley, Dr. George. *Bath Water.* Bath, England: Victor Morgan Books, Ltd., 1973.

Laas, William and Beicos, S.S., M.D. *The Water in Your Life.* New York: Popular Library, 1967.

La Riviere, Anne. "Environmental and Other Fears Fail to Kill Dreams of Harvesting Antarctic Icebergs." *Los Angeles Times,* 8 October 1977, Part I, p. 16.

Larson, T. E. "The Proposed Water Standards for Radium." *American Water Works Association Journal* (October 1976) 501-503.

May, Roger B. "French Bottler Tries to Replace U.S. Pop with a Natural Fizz." *Wall Street Journal,* 12 April 1978, p. 1.

McCure, Frank J. "Water Fluoridation, the Search and the Victory." U.S. Department of Health, Education and Welfare. Bethesda, Maryland: National Institute of Health and National Institute of Dental Research, 1970.

Merten, Ulrich, ed. *Desalination by Reverse Osmosis.* Cambridge: M.I.T. Press, 1966.

Moorman, J. J., M.D. *Mineral Springs of North America.* J. B. Lippincott and Co., 1873.

Muhler, Joseph C. and Hine, Maynard K., ed. *Fluorine and Dental Health —The Pharmacology and Toxicology of Fluorine.* Bloomington, Indiana: Indiana University Press, 1959.

Nagle, James J. "Getting Hooked on Bottled Water." *New York Times,* 4 March 1973, Part III, p. 3.

Noling, A. W. *Beverage Literature: A Bibliography.* Metuchen, New Jersey: The Scarecrow Press, 1971.

Price, Joseph M., M.D. *Coronaries, Cholesterol, Chlorine.* New York: Pyramid Books, 1971.

"Protecting Our Drinking Water," The EPA Journal, vol. 12, no. 7 (September 1986).

Ratliff, Rick. "Tapping Water." *Miami Herald,* 10 December 1976, Part AA, p. 1.

"Red Water in Bulgaria." *New York Times, 29 September 1935, Part III, p. 21.*

Schmeck, Harold M., Jr. "Study Finds Chemical Pollution of Drinking Water in 79 Cities." *New York Times,* 17 April 1975, Part I, p. 1.

Schroeder, Henry A., M.D. *The Trace Elements and Man, Some Positive and Negative Aspects.* Old Greenwich, Connecticut: The Devin-Adair Company, 1973.

Scully, Francis J., M.D. *Hot Springs, Arkansas, and Hot Springs National Park.* Little Rock: Pioneer Press, 1966.

Shuval, Hillel I., ed. *Developments in Water Quality Research.* Ann Arbor, Michigan: Ann Arbor-Humphrey Science Publishers, 1970.

Stone, William L. *Reminiscences of Saratoga and Ballston.* Virtue and Yorston, 1875.

Waring, Gerald A. *Thermal Springs of the United States and Other Countries of the World.* U.S. Department of the Interior, Geological Survey Professional Paper 492. Washington, D.C.: Government Printing Office, 1965.

Water Facts Consortium. "Glossary of Terms." Chicago: 1972.

Weiss, Harry B. and Kemble, Howard R. *They Took the Waters.* Trenton, New Jersey: The Past Times Press, 1962.

Wilkens, Emily. *Super Spas.* New York: Grosset and Dunlap, 1976.

Wilson, Eva D., Fisher, Katherine H. and Fuqua, Mary E. *Principles of Nutrition.* New York: John Wiley & Sons, 1975.

Yeo, Isaac Burney. *The Therapeutics of Mineral Springs and Climates.* New York: Cassells, 1904.

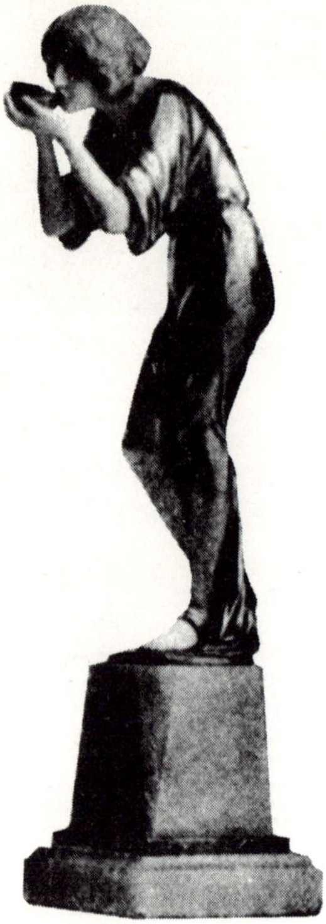